LIBRARIES AND LIBRARIANSHIP:
AN INTERNATIONAL PERSPECTIVE
Edited by Robert Stueart

1. *The Impact of Technology on Asian, African, and Middle Eastern Library Collections.* Edited by R. N. Sharma, 2006.

T0198020

The Impact of Technology on Asian, African, and Middle Eastern Library Collections

Edited by
R. N. Sharma

Libraries and Librarianship:
An International Perspective, No. 1

The Scarecrow Press, Inc.
Lanham, Maryland • Toronto • Oxford
2006

SCARECROW PRESS, INC.

Published in the United States of America
by Scarecrow Press, Inc.
A wholly owned subsidiary of
The Rowman & Littlefield Publishing Group, Inc.
4501 Forbes Boulevard, Suite 200, Lanham, Maryland 20706
www.scarecrowpress.com

PO Box 317
Oxford
OX2 9RU, UK

British Cataloging in Publication Information Available

Library of Congress Cataloging-in-Publication Data

The impact of technology on Asian, African, and Middle Eastern library
collections / edited by R.N. Sharma.
p. cm. — (Libraries and librarianship—an international perspective ; no. 1)
Includes bibliographical references and index.
ISBN-13: 978-0-8108-5448-2 (pbk. : alk. paper)
ISBN-10: 0-8108-5448-1 (pbk. : alk. paper)
1. Libraries—Information technology—Asia. 2. Libraries—Information
technology—Africa. 3. Libraries—Information technology—Arab countries.
4. Libraries—Information technology—Developing countries. 5. Information
resources—Asia. 6. Information resources—Africa. 7. Information resources—
Arab countries. 8. Information resources—Developing counties. 9. Libraries—
International cooperation. 10. Digital libraries. 11. Libraries—United States—
Special collections. I. Sharma, Ravindra N. II. Series.

Z674.83.A78I47 2006
025.00285—dc22

2006002784

For Mithlesh, Nalini, and Mohini

Contents

Introduction

Libraries have been part of the human civilization for centuries. They have played an important role in the development of people and societies. They have also been instrumental in storing and retrieving information for scholars and other users. Many changes have been introduced in libraries from time to time to meet the needs of the changing world, including the method of obtaining information from them. In ancient times, information was written and stored on clay tablets and handwritten materials, which changed to printed materials during the medieval times. Then came microforms, CD-ROMs, and the online storage method, including databases on the World Wide Web.

Technology is still very new to the libraries and their users. It was introduced in the Western world during the second half of the twentieth century, followed by a few poor nations in the Third World. My travels to Asia, Africa, and the Middle East on library assignments showed me that libraries in those regions are still far behind in the twenty-first century compared to the United States and other developed countries. Therefore, I invited sixteen well-known scholars, library educators, and librarians from these regions to write about the impact of technology on library collections and services in their countries. Three authors from the United States have written about Asian, African, and Middle Eastern collections in the United States and the impact of technology on these collections and services. Together they provide a clear picture of Third World libraries compared to those in the United States and of the ways to bridge the gap between developed and developing countries of these regions.

Ching-chih Chen has written about the Global Memory Net, a digital library project, and how it can help users to retrieve priceless resources from the global network, with many excellent examples from Asia, Africa, and the Middle East.

The book has five chapters on Asia, including one featuring the Online Computer Library Center, "OCLC Library Information Services and Libraries in Asia and the Pacific Region," written by Andrew H. Wang. This essay gives an overview of OCLC and its services with the help of technology and an emphasis on Asia and the Pacific Region. T. A. V. Murthy and V. S. Cholin discuss "The Impact of Technology on University Libraries in India" in their paper. Binh P. Le considers the case of Vietnam in his essay. He has illuminated the conditions facing libraries in the country thus far in the twenty-first century, marking their lack of technological resources. He also mentions the need of a proactive role by libraries to cultivate and nurture the community.

Jing Liu and Ian Yiliang Song discuss "The Impact of Technology on Chinese Library Collections and Services." They trace the history of the development of Chinese libraries and deal with technological developments and their application. The problems and challenges in Chinese libraries, based on the authors' own observations in China as well as literature reviews, are also discussed. Rajwant Chilana's subject is IT and the development of South Asian collections and services in the United States, as well as contributions of many organizations and committees in the development of collections and services.

There are five essays on Africa in the book. J. J. Britz and P. J. Lor discuss "The Role of Libraries in Combating Information Poverty in Africa." They observe that African libraries lack the required skills and abilities to obtain access to information and to interpret and apply it in a meaningful way for their users. The authors emphasize the important role libraries have to play in combating information poverty in Africa, including South Africa. Kay Raseroka reviews the challenges facing academic libraries in sub-Saharan Africa in the age of information. She highlights the factors that have led to the gradual introduction of technology in libraries, with specific reference to Botswana.

Three authors from Nigeria, C. C. Aguolu, I. Haruna, and I. E. Aguolu, write about "The Impact of Technology on Library Collections and Services in Nigeria." Their chapter deals with the applications of modern technologies and identifies the major technologies relevant to libraries. The authors also comment on the problems associated with the use of computer and related technology and make recommendations to improve the present situation. James J. Natsis writes about a unique partnership between the National University of Benin Library in West Africa and the West Virginia State University Library in Institute, West Virginia. The partnership was made possible with the help of a grant from the United Negro College Fund and the U.S. Agency for International Development. Gregory A. Finnegan shares his views on "The Impact of Information Technology on Africana Scholarship and Library Collections in the United States." He discusses modern information tools, including e-mail, the Internet, and the availability of full-text articles, books, and other documents. In his view, technology has brought scholars from all parts of the world together to contribute and share their ideas for the benefit of all users.

There are three essays on the Middle East. Mohammed M. Aman gives an overview of "The Impact of Technology on Libraries and Collections in the Arab Countries of the Middle East and North Africa." The progress in this region has been slow, but some steps have been taken to introduce technology in the libraries of the Arab world. They will help to shape the future of the region and offer Arab library users an opportunity to learn and use global library and information systems and databases. Sherif Kamel Shaheen's chapter, "Information Technology Applications in Information Work in Egypt," mentions Egyptian efforts to build the information society and the introduction of IT in library and information science research. Shaheen also deals with the problems and lack of proper facilities to make progress and apply universal standards customized and tailored to Egyptian needs, as well as the alarmingly high rate of information illiteracy in the country. Ali Houissa has written about "The Impact of Technology on Middle Eastern Collections and Services in the United States." He focuses on the high-speed communication and computer processing capabilities in the United States, which have helped information gathering and dissemination in Middle Eastern studies. The chapter includes a brief history of Middle Eastern collections in the country and efforts to benefit all users by the cooperative sharing of resources.

My contribution identifies major "Barriers in Introducing Information Technology in Libraries," specifically in Asia, Africa, and the Middle East. I suggest several solutions to the growing problem of bridging the gap between developed and developing countries and of achieving equality among users of libraries in the twenty-first century.

It has indeed been an excellent experience to work with all contributors. I want to thank each of them for accepting my invitation to write the chapters for this book and sending their contributions on time. The editorial staff at Scarecrow Press was very helpful in giving finishing touches to the manuscript and preparing it for publication. My thanks to the editors and staff for their hard work and to Mohini Sharma, my daughter, who worked on the manuscript and spent many hours to correct the problems in the essays and notes to prepare it to the publisher's specifications. Without her hard work and dedication in preparing the manuscript, it would have been difficult to complete the project. Finally, my wife, Mithlesh, and daughters Nalini and Mohini also deserve a special thanks for their support and patience while I was busy preparing and editing the manuscript. It is my hope that the chapters written by distinguished scholars for this book will give a clear picture of the impact of technology on library collections and services in Asia, Africa, and the Middle East, as well as in the United States, to all scholars, librarians, library educators, students, and other interested readers.

R. N. Sharma

I

WORLDWIDE

Chapter 1

The World's Treasure Is One Click Away at the Global Memory Net

Ching-chih Chen

ABSTRACT

Global Memory Net (GMNet), expanded from Chinese Memory Net (CMNet), is a digital library project supported by the International Digital Library Program of the U.S. National Science Foundation. It deals with multimedia resources but currently focuses on digital images. Although "memory" can be worthy information related to all subject areas, currently it contains mainly global cultural, historical, and heritage multimedia contents that build on the multimedia resources related to the first emperor of China's terracotta warriors and horses collected for Project Emperor–I with the support of the National Endowment for the Humanities (NEH) in the mid-1980s. This chapter describes what GMNet is, what cutting-edge technologies were used, how it can deliver the world's treasures at a simple click of the mouse, and how it can link invaluable global resources from distributed systems over the global network.

INTRODUCTION

My early interactive videodisc R&D work with the Project Emperor–I in the mid-1980s and later the multimedia R&D related to the first emperor of China's famous terracotta warriors and horses in Xian received considerable media attention and produced a long list of publications, of which two are cited here. Project Emperor–I demonstrated how multimedia

3

technology could change the way we seek, demand, and use information. Two decades have passed since then, and in technological terms, two decades are very long time! The information technology innovations have intertwined with the interdisciplinary knowledge base, which is propelling the twenty-first century's knowledge economy. By the end of 2004, use of multimedia had not only become the mainstream practices but was taken for granted. In fact, fueled by enormous progress in science and technology, we have come a very long way from the use of interactive multimedia technology in the workstation environment to the global networked environment. We have moved from the use of hardcopy and analog resources to digital content, which users can search, retrieve, and use instantly to meet their needs over the global network with no national boundaries. We have also moved from offering the world multimedia content on one specific subject to offering the digital content of all media formats on all subjects, instantly. We are truly living in a new period of unprecedented opportunities and challenges! So, in this digital era, we have witnessed the exciting convergence of content, technology, and global collaboration in the development of digital libraries with great potential for providing universal information access.

Thus, today's information seekers, whether they are the general public, schoolchildren, or those from research and higher education communities and whether they are seeking information for education, research, entertainment, or enrichment, seek needed information in very different ways from before. From the information resources point of view, the old model of "owning" a collection has given way to "sharing," and the new emphases have shifted from possessing large physical libraries, valued for their large numbers of volumes, to "virtual libraries" digitally distributed all over the world.

In the last two decades, I have experienced many of these transformations "up close and personal" through my own R&D activities—from the creation of interactive videodisc and multimedia CDs in the 1980s and 1990s to leading a current international digital library project, Global Memory Net, supported by the International Digital Library Program of the U.S. National Science Foundation.

WHAT IS GLOBAL MEMORY NET?

From Project Emperor–I to Chinese Memory Net

In the early 1980s, the by-product of Project Emperor–I was a set of interactive videodiscs, called *The First Emperor of China*, the content of which later was converted to a popular multimedia CD of the same title in 1991 and published by the Voyager Company. This NEH project has resulted in thousands and thousands of invaluable images and multiple hours of videos of incredible value to scholars and general citizens. The U.S. National Science Foundation initiated and funded the first phase of Digital Libraries Initiatives (DL-I) in the first half of 1990s, and then the Second Phase of Digital Libraries Initiatives (DL-II). The funded NSF projects have developed some exciting digital library technologies, but the contents of digital libraries remain limited. This has motivated us to continue our efforts in building more image and other multimedia contents after Project Emperor–I and develop more complete descriptive information (metadata) of the image resources. Our effort paid off. In 1999, when NSF introduced its International Digital Library Program (NSF/IDLP), we proposed Chinese Memory Net (CMNet) and it became one of the first NSF/IDLP Projects.

The NSF's support of CMNet since 2000 is intended to develop a model for international collaboration with various R&D activities in digital libraries. It hopes to accomplish "more" with "less," avoid duplication efforts, and capitalize R&D results from other digital library R&D projects that receive major funding. Thus, extensive efforts were made to develop collaborative infrastructure with collaborators in:

- Beijing—Peking University and Tsinghua University;
- Shanghai—Shanghai Xiao-tong University;
- Taipei—Academia Sinica, National Taiwan University, and National Tsinghua University; and
- The United States—Carnegie Mellon University in Pittsburgh, and Penn State University.

In the short four years, it has made true progress in developing collaborative infrastructure for digital library development. Both CMNet and the New Information Technology (NIT) 2001 conference in Beijing, organized by this author, played an important role in promoting the development of digital libraries in China and partially in Taiwan. CMNet's core content builds upon the large quantity of visual materials of the earlier interactive videodisc (1986) and multimedia CD (1991) products, *The First Emperor of China*. For each image included, extensive research efforts were made to provide relevant descriptive data (metadata) with annotations, as well as links to relevant references and texts whenever possible.

Extensive Metadata Preparation and Content Development

While building the digital library community and infrastructure, CMNet also started the labor-intensive R&D activity in content and

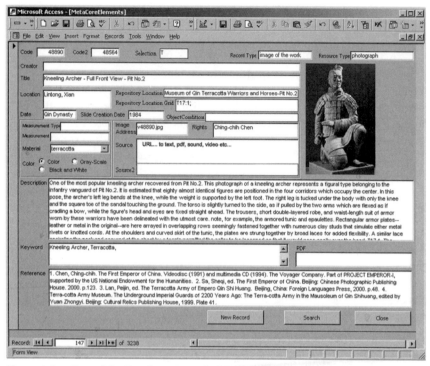

Figure 1.1.   One of the Database Records for an Image of a Kneeling Archer

metadata building. Leveraging on the extensive descriptive annotations of the earlier interactive videodisc project, more effort has been made since 2000 for the labor-intensive but much needed metadata work using Microsoft Access to create the backend database. Figure 1.1 shows one of the database records for an image of a kneeling archer. One can imagine the enormous effort spent. With these data, traditional searches using each of the metadata fields can be made easily. This forms the basis for the "traditional" search of Global Memory Net to be described later. Figure 1.2 shows the quick results when the "head" of the "title" field is searched.

This tedious activity has paid off because these invaluable image resources and metadata have formed an attractive basis for a number of exciting and productive technology-oriented collaborations with several computer scientists both in the United States and elsewhere, who badly need relevant real-life data to test their developed technologies. Some of the collaborative research activities are listed in the following, with more elaboration on some of the activities in the latter part of this chapter:

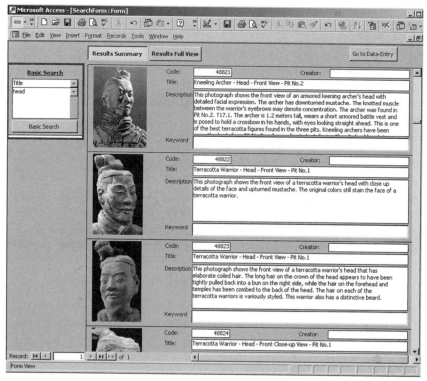

Figure 1.2.   The Quick Results When the "Head" of the "Title" Field Is Searched

- Open Archive Initiative (OAI) research—with the collaborators in China, since each is using quite different metadata
- Intelligent agent and text-based image retrieval—in collaboration with Prof. V. W. Soo of the National Tsinghua University in Hsinchu, Taiwan
- Semantic sensitive content-based image retrieval—in collaboration with Prof. James Z. Wang of Penn State University
- Digital video using the Informedia technologies—in collaboration with Prof. Howard Wactlar of Carnegie Mellon University
- Machine learning for annotation—in collaboration with Profs. James Z. Wang and Jia Li of Penn State University.

From Chinese Memory Net to Global Memory Net (GMNet)

Once it is possible to develop a multimedia digital library in one discipline or for one geographical area, it is upwardly scalable to include more subject topics and bigger geographical areas. This has been the case with

Figure 1.3.   Topics Included on GMNet

the expansion of the scope of CMNet to GMNet since 2002. CMNet project has concentrated on images and video related to China's culture, history, and heritage. Yet, in the last two years, the use of cutting-edge technologies in the organization and retrieval of multimedia contents, specifically the digital images, have attracted collaborative possibilities with several major institutions in different countries other than China. Figure 1.3 shows only a very small portion of topics included. It is clear that most of them don't belong to CMNet. The need to expand the project's coverage and scope to Global Memory Net becomes obvious. Now, GMNet can cover the "memory" of any part of the globe and thus have the flexibility to cover valuable contents of more than two hundred countries in the world as long as digital contents are available.

Expanding CMNet to GMNet expedites the digital library collaborative development and frees the R&D activity from unnecessary logistical delays and inflexibilities. With over two hundred countries in the world, there are endless opportunities for digital collection development, digital partnership, and collaborative technical research activities. In addition, the current direction also widens the possibilities of GMNet serving as a functional multimedia gateway or portal to the world's invaluable "memory" resources in all types of resource organizations—libraries, museums, archives, academic institutions, and others. This offers incredible opportunities for easy universal access to the world's treasures. GMNet offers users the world instantly!

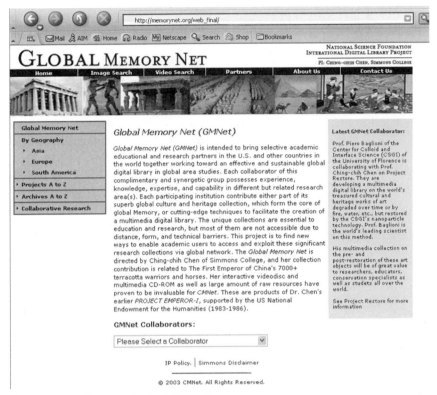

Figure 1.4.   A Tentative GMNet Home Page

WHAT DOES GLOBAL MEMORY NET COVER?

The Scope of Global Memory Net

The name of Global Memory Net clearly articulates the potential coverage and scope of this project, and let's state again those key elements that are clear indicators of the scope of GMNet:

1. Global—*G* stands for global coverage, as shown below in figure 1.4, a tentative GMNet home page. It shows clearly that there is a space holder for all countries in the world even though this tentative homepage currently has listed only a few continents and countries under each in the "geographical" category. The reality is that all continents will be included, and under each continent, all the countries can be listed. A plan is being made to provide a world map to permit users to access to any country or area by clicking on the proper location of the map.

2. Memory—*M* for "Memory" refers to all the the treasures that GMNet has the structure to cover— all kinds of invaluable memories related to culture, heritage, history, art, music, science, technology, medicine, and more. However, at this initial stage, and with initial entry of the extensive visual memory related the first emperor of China's terracotta warriors and horses, GMNet focuses on globally significant cultural, historical, and heritage materials. Once work related to this aspect is well under way, GMNet will expand to cover other kinds of "memories." Currently the valuable image collections can be accessed by "Geography," "Project," "Archive," and so on.

3. *Net*—"Net" means that GMNet hopes to network all significant world resources together. Instead of encouraging the development of small and fragmented digital libraries, it hopes to be a networked portal to offer needed resources instantly with the simple click of the mouse. It is clear now that GMNet should not try to have the world's valuable "memory" collections in one location. Instead, it should try to bring distributed collections around the world together. Thus, from the few sample images, GMNet can link the users to the desired image collection instantly.

How Global Memory Net Offers the World!

GMNet supplements well another major digital project, the China-US Million Book Digital Library Project (hereafter referred to as the Million Project), of which Dr. Raj Reddy and this author are U.S. coprincipal investigators. The U.S.-NSF's Million Project collaborates with over a dozen Chinese higher education and research institutions whose support comes from the Chinese Ministry of Education.

The Million Project is a major digitization project. It is currently mainly text-based.[1] GMNet, however, concentrates on multimedia resources— images, videos, music, and other audio resources. Currently, it is mainly related to digital image resources. For this reason, most of the the following discussion about GMNet offer examples of cutting-edge content-based image retrieval of digital images, with only limited mention of digital video activity. GMNet's current valuable image collections cover many countries in the world, as shown in figure 1.4. New subjects are being added constantly.

For images of the first emperor of China's terracotta warriors and horses and those collections with a large number of images, GMNet is a comprehensive image digital library. For users of small collections of cultural and heritage images included in GMNet, GMNet serves as an effective digital portal offering the world instantly to the information seekers. Once a user selects the desired images from those retrieved, he/she can then be referred to the relevant site directly for more information. More discussion on how the images collections are accessed and retrieved is provided in the following.

Image Retrieval

The simplest way describe all the features of GMNet is just to imagine taking a visual tour of a selected cultural, heritage, or historical topic while sitting at one's computer. This soon-to-be-available GMNet on the Internet will provide image retrieval capabilities with considerable textual support, in a way not possible before. For example, from the page like that shown in figure 1.4, if one selects the *Emperor* collection, one can go to "China" and then "Emperor Image Base" quickly. Then one will be able to retrieve invaluable images related to the first emperor of China by conducting either the traditional search (using Google protocol if predefined specifics of the images are known), or by the cutting-edge semantically sensitive content-based image retrieval.

Traditional Image Search

In the traditional search, one can search literally every field of the metadata, such as creator, title, location, time period, description, keyword, reference source, and so on by using the Google syntax. In this approach, keyword search is likely to be the most popular one. If "keyword" search is selected, the search terms are typed using the Google syntax in which a "+" indicates the "required" term. For example, if one would like to find the images related to the end of tiles, one will type (+tile, +end) (i.e., both "tile" and "end" are required to appear as keywords); then, almost instantly, from the thousands of images in the image base, the screen (see figure 1.5) will show the first fifteen tiles ends located. This type of search, enables precise retrieval of available images.

Semantically Sensitive, Content-Based Image Retrieval

In most cases, one generally does not have any idea on what kind of images are available in GMNet except that it is international in coverage. Just as in a library, we need to provide the users an opportunity to browse the stack and find what they need and want. Yet, most image databases do not offer users the chance to browse. "Browsing" and "random showing" capabilities are the special features of GMNet. Here, we use the cutting-edge content-based image retrieval technique, SIMPLIcity, developed at Stanford University under NSF's DL-I phase, and then at the Penn State University under NSF/ITR funding. This allows users

Figure 1.5.   The Screen Indicating the First Fifteen Tiles Ends Located

to browse, retrieve, enjoy, and learn in just seconds through multiple thousands of digital images. As shown in figure 1.6, under the "Traditional Search" on the left panel, there are buttons that offer the user three possibilities.

The user chooses from:

- Random—By clicking on this, images in the image base will show up randomly
- Browse—By clicking on this, users will be able to browse images ten or fifteen at a time from page to page until they spot the desired image
- URL—The user will be able to ask the system to find images that are similar to the one located on the given URL address.

Until recently, most archival images were not available in digital form. Now we have together in one place a large quantity of invaluable digital materials from multiple countries. One can ask the system to bring out image icons randomly, or browse the images by displayed icons page to page until one locates the image of interest. For example, when the icons of the mages of the *Emperor* collection are displayed randomly (see figure

Figure 1.6. The Three Buttons under the "Traditional Search" on the Left Panel That Offer Users Three Possibilities

1.7), the user spots a map image of particular interest. In this case, the user can simply ask the system to retrieve all images similar to the one chosen by simply clicking "Similar" without typing any word on the keyboard. GMNet will display in seconds all the maps in the collection similar to the one selected (see figure 1.8). This opens up all possibilities for all related maps that were totally unknown to the user prior to the showing, thus truly enhancing the learning experience of the user.

Once these massive numbers of images are displayed, one will be able to find instantly more textual descriptive information as well as reference sources and, in some case, full-text descriptions on a chosen map by clicking "Info" (to be elaborated further in next section). If the chosen image needs to be enlarged, then click on "larger," and multiple levels of zooming will be possible to show the desired details of the map or a portion of the map. Concurrently, a dynamic digital watermark will be instantly generated at any zooming level to offer the "ownership" information of the image as shown in the bottom right corner. This overcomes the problem and worry related to copyright and intellectual property.

Figure 1.7.   The Icons of the Images of the *Emperor* Collection Displayed Randomly

Example: When the icons of the images of the *Emperor* collection are displayed randomly (see figure 1.9), one spots the image related to "Han silk" of interest. In this case, one can ask the system to provide "similar" images by clicking "Similar" without typing any word. GM-Net will display in seconds all the images on "Han silk" in the collection similar to the one selected (see figure 1.10). From the massive numbers of images displayed, one can enlarge a chosen image by clicking on "larger " and the enlarged image with dynamic digital water mark will be instantly generated to offer the "ownership" information of the image as shown in figure 1.11. One can also find more textual descriptive information as well as reference sources and, in some cases, the full-text original source on a chosen image instantly by clicking "Info" as shown in figure 1.12.

The examples shown above should reveal vividly the potential of GMNet for delivering invaluable "memory" image contents to the world. Needless to say, this similar technique can be used to retrieve any image collections available for inclusion in GMNet, regardless of subject.

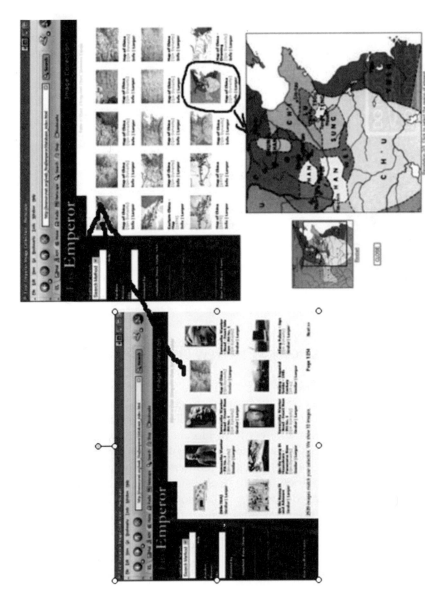

Figure 1.8.   GMNet Displaying All the Maps in the Collection Similar to the One Selected

Figure 1.9. The Icons of the Images of the *Emperor* Collection Are Displayed Randomly and One Spots the Image Related to "Han Silk" of Interest

GLOBAL MEMORY NET—NEW COLLABORATION

The move from the Chinese Memory Net to GMNet has enabled us to expand our collaborative community and subjects greatly, far more than we could imagine during the first three years of CMNet. It began with our exciting collaboration with University of Florence on Project Restore with the exciting pre- and post-restoration images of the invaluable Italian art objects (see figure 1.13) using the world-renowned nanoparticle chemistry technology. This technology for physical restoration of damaged art objects developed by our collaborator, Professor Piero Baglioni and his research group at the Center for Colloid and Interface Science (CSGI) of the University of Florence, should have great potential for the restoration of all kinds of artifacts in different parts of the world. In fact, as Professor Baglioni and colleagues have reported, many world-leading museums are collaborating with him in using his technology to restore their artifacts and manuscripts.

In addition to projects like Project Restore, other collaborative collections have mushroomed to include many countries and subjects. Collec-

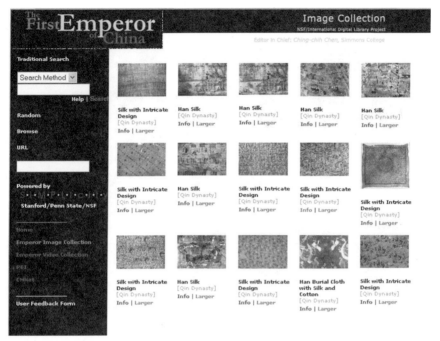

Figure 1.10. By Clicking "Similar" without Typing Any Word, GMNet Will Display All the Images on "Han Silk" in the Collection Similar to the One Selected

tions related to "living museums" around the world in open space, such as historical sites in China, India, Egypt, Greece, Italy, and other countries, are abundant, notably:

- China—Chinese painting, historical unique collections, historical site, architecture, etc.
- Cambodia—Ancient temples, etc.
- India—Architecture, palaces, temples, goddess, etc.
- Vietnam—Historical development of the former Saigon
- Italy—Historical artifacts, art objects
- Europe—Cathedrals, castles, ancient cities, etc.
- World—Global musical instruments
- World digital collections, national libraries, etc.

In addition to these, collaborations with major national libraries and international organizations are beginning to surface. For example, under the leadership of Dr. Hwa-wei Lee, the head of the Asian Division of the Library of Congress, we have begun to explore an exciting collaboration

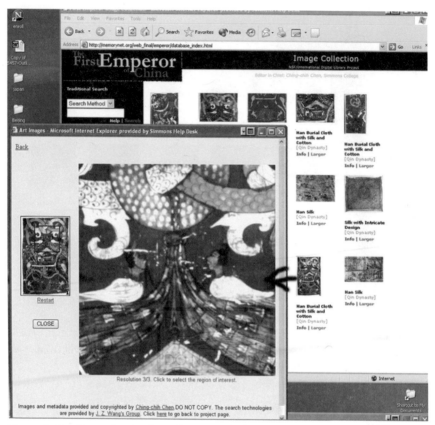

Figure 1.11.   From the Number of Images Displayed, One Can Enlarge a Chosen Image by Clicking on "Larger," and the Enlarged Image with Dynamic Digital Watermark Will Be Instantly Generated to Offer the "Ownership" Information of the Image

by including the unique Naxi manuscripts' images from the Library of Congress in GMNet (see figure 1.14). Differing from the traditional Web presentation of the collection as shown in figure 1.15 on the Library of Congress site, where images are shown one at a time, GMNet is able to retrieve and show all related similar images all at once, as shown in figure 1.14, and thus has boosted the accessibility and value of the Naxi manuscripts' images.

Figure 1.14 also shows that we are also exploring the collaboration with UNESCO's Memory of the World Programme. We have identified over 1,400 digital collections in the world, and it is possible for us to retrieve all websites of digital collections of similar color and design of an organization instantly, such as those of the UNESCO's Memory of the World Pro-

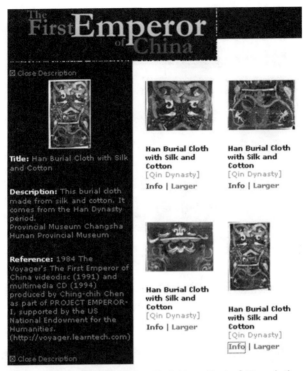

Figure 1.12. One Can Also Find More Textual Descriptive Information as Well as Reference Surces and, in Some Cases, the Full-Text Original Source on a Chosen Image Instantly by Clicking "Info"

gramme as shown in the third row of figure 1.14, with all Web pages in "yellow" but with different contents. Once a website is selected, information on the site can be located instantly, and the user can be linked to the site instantly as well. Currently, UNESCO has ninety-one digital collections from forty-five countries registered under Memory of the World; thus, our digital portal has certainly boosted the accessibility and value of these collections instantly.

FUTURE DEVELOPMENT

In addition to continue the building of a great variety of image collections and global partnership, future development will move more aggressively to the areas of digital video, sound, and audios.

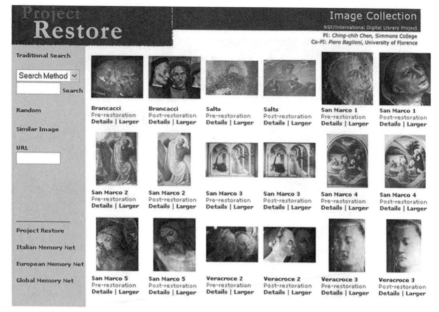

Figure 1.13. The Exciting Pre- and Post-Restoration Images of the Invaluable Italian Art Objects

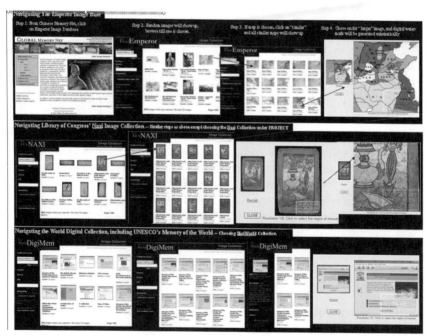

Figure 1.14. The Unique Naxi Manuscripts' Images from the Library of Congress in GMNet

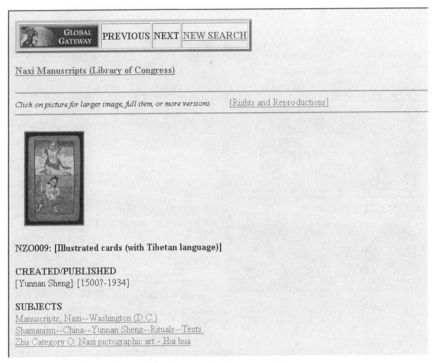

Figure 1.15.   The Traditional Web Presentation of the Collection

Digital Video and Informedia Technologies

Carnegie Mellon University's well-known Informedia Project is one of the six original NSF/DLI-1 projects. It has continued its development in digital video–related technologies and tools ever since 1995. Collaboration between Informedia and CMNet since 2000 has enhanced perspectives from cultural and historical video documentaries. Its multilingual (English and Chinese) aspect has also posed challenges in its speech recognition–related research. When the Informedia technology is ready for Web-based use, GMNet will be ready to use it. Figure 1.16 shows some of the screens generated from the latest collaboration. The upper left shows that when "emperor" is searched, sixty video segments with that word have been identified and the six most relevant ones are retrieved and shown in the left middle screen; these segments can be visualized in a time line as shown in the lower left screen. The map is shown in the upper right screen, and when one of the videos is chosen, the video will play in the upper right of the lower right screen, and below that, the actual text will also be displayed with the word "emperor" highlighted in red. The running bar between the video segment and the textual annotation shows the red

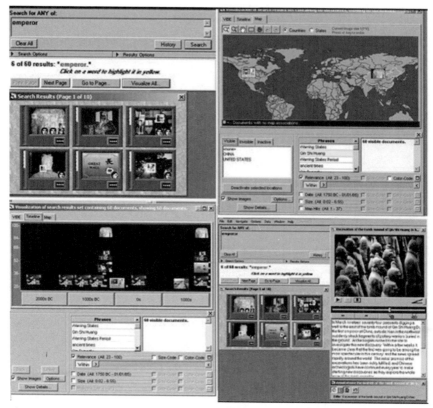

Figure 1.16.   Some of the Screens Generated from the Latest Collaboration

line(s) where the word "emperor" will appear when the video playing reaches the indicated area(s). This is an incredible showing of how cutting-edge technologies can be used to enhance the retrieval of digital videos in a way not possible before.

As mentioned earlier, although GMNet has concentrated thus far on digital cultural and heritage image collections, we are beginning to explore collaborative possibilities in other multimedia formats and multilingual aspects. In addition to the possibility of using Infomedia technologies for the retrieval of digital videos, we will provide the "traditional" ways of searching digital videos as well.

Digital Voice and Music

In addition to digital videos, our research will also explore the potential use of sound and music. One of the perfect starting points will be with the world's musical instruments. Figure 1.17 shows that such an image base

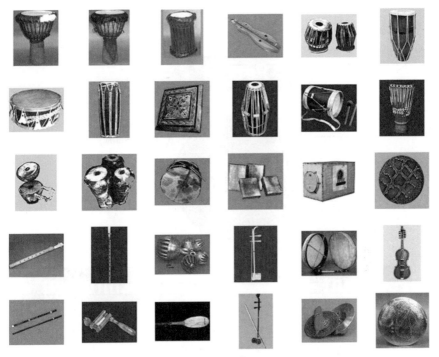

Figure 1.17.   An Image Base Being Constructed

is being constructed. It is our hope that the instruments will also be linked to music and sound when available. This is another exciting collaborative area with great potential! Another possible area would be with the language learning and writing.

FURTHER IMAGE MANAGEMENT AND
RETRIEVAL TECHNIQUES

One final mention of an exciting new activity would have to be our newly funded NSF/IDLP [NSF/IIS-Special Projects (IIS)] two-year project from 2004–2006, entitled "International Collaboration to Advance User-oriented Technologies for Managing and Distributing Images in Digital Libraries" with James Z. Wang of Penn State University and Jianbo Shi of University of Pennsylvania as co-PIs. This project will develop user-oriented image management of distribution technologies for digital libraries. An interdisciplinary team of computer and information scientists from the United States, China, and Taiwan will investigate efficient ways to search digital collections of images using an integrated approach. The

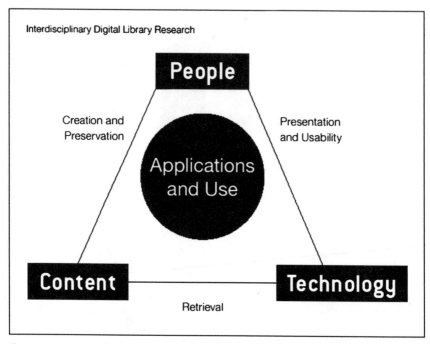

Figure 1.18.   Interdisciplinary Digital Library Research

team will use real-world digital library datasets to develop user-oriented technologies suitable for practical deployment. Notably, the research will utilize an existing collection consisting of a large quantity of images associated with the First Emperor of China's terracotta warriors and horses at all levels of resolution. This research will also capitalize the existing rich descriptive information for research purposes. In addition to ontology-based image retrieval, the project will deal with machine learning–based and content-based image retrievals, as well as the difficult object–based partial image searches. We also hope to extend the intellectual property (IP) protection technique. For more information on the project before the website is available for public use, visit www.memorynet.org

CONCLUSION

During 1997–2002, I was privileged to serve on the U.S. President's Information Technology Advisory Committee (PITAC). Our PITAC's Digital Library Panel's Report, *Digital Libraries: Universal Access to Human Knowledge,* has a vision for digital libraries:

All citizens anywhere anytime can use any Internet-connected digital device to search all of human knowledge. . . . In this vision, no class-room, group, or person is ever isolated from the world's greatest knowledge resources.

This is a vision more easily said than done! There are many obstacles on the road, so we are a long way from approaching this "elusive" vision.

In considering international digital library research and development, it is important for us to revisit the conceptual model presented by the DE-LOS/NSF Working Group on Digital Imagery for Significant Historical, Cultural and Heritage Materials, of which I am a U.S. cochair (figure 1.18). From this model, it is clear that GMNet is developing substantial multimedia contents—currently mostly images—both in-house and through linkages with distributed systems through the use of the global network. The retrieval of these contents uses both conventional and cutting-edge technologies. They are made available for use by the general public as well as by scholars and researchers via the Web. This chapter clearly addresses mainly the "content" aspects, with mentions of what the technologies are used to do and to present. More detailed discussions on the technology can be found in the references.

As to "content," from "sharing" and "accessing" points of view, we must first have much more high-quality digital content, we must collaborate internationally in content building because no one can have everything, and then we must have the technology to cope with the contents, as well as the infrastructure to deliver, access, and retrieve them. This is what GMNet is inspired to do specifically in content building and method development areas. The new collaboration and new R&D activities have expanded our research horizon and have offered us great opportunities for digital library community building, for making digital collections alive and accessible, and for contemplating a much more practical R&D agenda in areas of metadata standards, interoperability, scalability, retrievability of difficult multimedia contents, and usability of these resources for knowledge creation.

It is gratifying that in a short couple of years, GMNet has demonstrated how international collaboration and community building in promoting large-scale content building, coupled with new technological tool and method development, can indeed offer users the world in a way not possible before. The potential for delivering and marketing invaluable world multimedia resources should also be clear. The best is yet to come!

ACKNOWLEDGMENTS

Chinese Memory Net and Global Memory Net are supported by the U.S. NSF/IDLP under grant no. IIS-9905833. Project Emperor–I was supported by the Humanities in Libraries Program of the U.S. National Endowment for the Humanities. Examples given on the use of SIMPLIcity are in collaboration with James Z. Wang of the Penn State University, whose work is supported by the NSF/ITR program under grant no. IIS-0219271. The new image management work is funded by the U.S. NSF/ CISE/IIS/ IDLP under grant no. IIS-0333036.

NOTES

This chapter combines contents and modifies them from many keynote and invited speeches given during 2004 by the author in different parts of the world, including Hanoi, New Delhi, Bangalore, Mysore, Dubrovnik, Orlando, Beijing, Guangzhou, Miami, Shanghai, Hainan, and Yokohama. While each talk contains a basic introduction to Global Memory Net, it also presents new information on latest developments of or new topical subjects covered by the National Science Foundation's International Digital Library Project. Duplication of information from previous publications is unavoidable.

1. *Million Project*'s Chinese partners include six Phase I institutions: Chinese Academy of Sciences (Northern center), Zhejiang University (Southern center), Peking University, Tsinghua University, Fudan University, and Nanking University; and Phase II institutions—Beijing Normal University, Jilin University, Shanghai Jiao-tong University, Zhong-san University, Wuhan University, Sichuwan University, and Xian Jiao-tong University.

BIBLIOGRAPHY

Ching-chih, Chen. 1991. *The First Emperor of China*. Santa Monica, CA: The Voyager Company. This is a videodisc product and is the popular version of this title [A Cindy Award videodisc product].
Ching-Chih, Chen. 1994. "Multimedia and the First Emperor of China: Moving toward a digital knowledge base," *Multimedia Today* 2, no. 2 (April 1994): 68–71.
Ching-chih, Chen. 2002. "Digital Library Development in the US and Global Reach: Roles of Federal Government Agencies and PITAC." Pp. 17–27 [Chinese version], 28–41 [English version] in *Proceedings of the Digital Library—IT Opportunities and Challenges in the New Millennium: An International Conference*, Beijing, July 9–11, 2002. Beijing, China: National Library of China. Pp. 17–27.
Ching-chih, Chen. 2004. "Global Memory Net: Digital Portal For Global Resource Sharing and Closing Digital Divide." Poster session presented at the 70th International Federation of Library Associations and Institutions General Conference, August 2004, Buenos Aires, Argentina.
Ching-chih, Chen. 2004. "Global Memory Net: New Collaboration, New Activities, and New Potentials," in *Proceedings of the International Asian Conference on Digital Libraries*. Shanghai, China: IACDL.

Ching-chih, Chen. 2004. "Global Memory Net Offers Users the World Instantly," *Library Times International* 21, no. 1 (July 2004): 1–4.

Ching-chih, Chen. 2004. "Global Memory Net Offers the World Instantly: Potentials for Universal Access to Invaluable Conten*ts*," *in Proceedings*, CCDL: Digital Library—Advance the Efficiency of Knowledge Utility, Beijing, September 5–8, 2004. Beijing: National Library of China.

Ching-chih, Chen. 2004. "Global Memory Net: Potential and Challenges for Archiving and Sharing Cultural and Heritage Resources." Pp. 3–10 in *Proceedings of the ICDL International Conference on Digital Libraries 2004*, Delhi, February 25–27, 2004. Delhi, India: TERI.

Ching-chih, Chen. 2004. "Past Forward—Digital Media for Cultural Heritage: The case of the Global Memory Net." Invited lecture given at the 10th Annual Lecture of Informatics, Bangalore, India, February 29, 2004.

Ching-chih, Chen. 2004. "The Promise of International Digital Library Collaboration for Innovative Use of Invaluable Resources." Pp. 7–15 in *Human Information Behaviour & Competences for Digital Libraries*. Keynote in *Proceedings*, The Libraries in the Digital Age, May 25–29, 2004, Dubrovnik and Mljet.

Ching-chih, Chen. 2004. "Sharing the Abundant Human Knowledge When Old Cultures Meet in Cyberspace: Global Memory Net for International Collaboration," *U&I Special Issue* (Bangalore, India: March–April 2004), pp. 4–5.

Ching-chih, Chen, ed. 2001. *Global Digital Library Development in the New Millennium: Fertile Ground for Distributed Cross-Disciplinary Collaboration*. Beijing: Tsinghua University Press.

Ching-chih, Chen, and James Z. Wang. 2002. "Large-scale Emperor Digital Library and Semantics Sensitive Region-based Retrieval." Pp. 454–62 in *Proceedings of the Digital Library—IT Opportunities and Challenges in the New Millennium: International Conference*, Beijing, July 9–11, 2002. Beijing, China: National Library of China.

Ching-chih, Chen, and Kevin Kiernan, eds. 2002. *Report of the DELOS-NSF Working Group on Digital Imagery for Significant Cultural and Historical Materials* (2002).

A Collection of Hu Boxiang's Painting: Memory of Hu Boxiang's Centennary Birthday Celebration. 1997. Shanghai, China: Shanghai Publishing House of Calligraphy and Painting.

James Z. Wang, Kurt Grieb, Ya Zhang, Chen Ching-chih, Yixin Chen, and Jia Li. Forthcoming. "Machine Annotation and Retrieval for Digital Imagery of Historical Materials." *International Journal on Digital Libraries*, Special Issue on Multimedia Contents.

Piero Baglioni, Rodorico Giorgi, and Ching-chih Chen. 2003. "Nanoparticle Technology Saves Cultural Relics: Potential For a Multimedia Digital Library." Paper presented at the DELOS/NSF Workshop on Multimedia in Digital Libraries, Chania, Crete, Greece, June 2003. memorynet.org/pdf/baglioni_crete.pdf/.

Soo, V. W., C. Y. Lee, C. C. Yeh, and Chen, Ching-chih. 2002. "Using Sharable Ontology to Retrieve Historical Images." Pp. 197–98 in *ACM/IEEE JCDL Proceedings*. Portland, Oregon. 2002.

Soo, V. W., C. Y. Lee, C. C. Lin, S. L. Chen, and Chen, Ching-chih. 2003. "Automated Semantic Annotation and Retrieval Based on Sharable Ontology and Case-based Learning Techniques." Pp. 1–12 of paper presented at the ACM/IEEE Joint Conference of Digital Libraries, Houston, Texas, May 2003.

U.S. President's Information Technology Advisory Committee. 2001. Digital Library Panel, *Digital Libraries: Universal Access to Human Knowledge*. Washington, DC: "PITAC, 2001 (accessed December 15, 2004).

Wang, James Z., Jia Li, and Chen, Ching-chih. 2002. "Interdisciplinary Research to Advance Digital Imagery Indexing and Retrieval Technologies for Asian Art and Cultural Heritage." Pp. 1–6 in *Proceedings of the ACM Multimedia, Workshop on Multimedia Information Retrieval*, Juan Les Pins, France, December 2002. New York: Association for Computing Machinery.

Wactlar, Howard D., and Chen, Ching-chih. 2002. "Enhanced Perspectives for Historical and Cultural Documentaries Using Informedia Technologies." Pp. 338–39 in *Proceedings of the ACM/IEEE Joint Conference of Digital Libraries*, Portland, Oregon, July 15–18, 2002. New York: Association for Computing Machinery.

II

ASIA

Chapter 2

OCLC Library Information Services and Libraries in Asia and the Pacific Region

Andrew H. Wang

ABSTRACT

Founded in 1967 and headquartered in Dublin, Ohio, OCLC Online Computer Library Center, Inc., a nonprofit, membership library information service and research organization, is "the United Nations of libraries." OCLC operates the largest library information network of libraries in the world, providing online library information services to fifty-three thousand institutions in 108 countries and territories. OCLC was formed to establish, maintain, and operate a computerized library network and to promote the evolution of library use, of libraries themselves, and of librarianship, and to provide processes and products for the benefit of library users and libraries. OCLC's vision is to become the leading global library cooperative, helping libraries serve people by providing economical access to knowledge through innovation and collaboration.

WorldCat (the OCLC Online Union Catalog) is the global union catalog of 10,000 libraries around the world, and thus the largest database of bibliographic and holding information in the world. Cooperative cataloging is OCLC's flagship service. Cataloging on WorldCat is not merely a matter of finding and copying MARC records but, rather, of internationalizing a library's collection, participating in the sharing of resources, and linking library patrons through the World Wide Web via Google and Yahoo! OCLC's FirstSearch Service provides online access to more than seventy databases and is the most popular online information service among university libraries in the United States. Archival access to e-journals after libraries have cancelled subscriptions to them is OCLC's unique and revolutionary guarantee to libraries. OCLC netLibrary eBooks was the pioneer of online access to e-book and the leading e-book provider in the world.

Established in 1986, OCLC Asia Pacific provides online information service to 3,600 libraries in eighteen countries and territories in Asia and the Pacific region. OCLC Asia Pacific also works with many key library consortia in the region. Aside from delivering information to the region, libraries in the region also contribute information of Asian content to WorldCat for sharing of the information globally.

OCLC: AN INTRODUCTION

OCLC Online Computer Library Center, Inc., a nonprofit, membership library information service and research organization, is "the United Nations of libraries." OCLC operates the largest library information network of libraries in the world, providing online library information services to 53,000 institutions in 108 countries and territories. OCLC is not a commercial company but, rather, an organization of libraries, similar, to some extent, to library associations. The fundamental objective of OCLC is to solve common problems of libraries through library cooperation and collaboration built upon information technology. No libraries in the world can be self-sufficient or do everything alone. OCLC has built not only the largest information network of libraries but also a people network of librarians on a global scale upon which the spirit of cooperation and collaboration is built and maintained.

Headquartered in Dublin, Ohio, OCLC was founded in 1967 by the presidents of fifty-four private and state colleges and universities in the state of Ohio as a nonprofit membership organization to help libraries, through cooperation and sharing of information and knowledge resources, gain further access to the world's information in support of education and research, to reduce the rate of rise of the unit cost of the information and knowledge, and to conduct ongoing research to develop technologies to support that mission. OCLC has grown from a library cooperative of fifty-four academic libraries to being a global library cooperative today. No other information provider in the world invests as much in research and development of library information and library information technology as does OCLC. The information services that OCLC provides are to support education and research. OCLC does not provide information that is in the nature of entertainment.

Being a nonprofit organization, however, does not mean that OCLC services are free of charge, nor does it mean that OCLC must lose money. OCLC is not a government agency, and therefore does not receive funding support from any government. OCLC is not a grant foundation, nor is it a charity organization. In order to recover operating costs, OCLC must charge libraries for using the information services. OCLC's revenue must be equal to or more than its operating costs, as OCLC must be self-supporting financially. The nonprofit status of OCLC means that OCLC exists for a public good and thus is exempt from paying income taxes. Besides, being a nonprofit organization, OCLC cannot issue stocks and therefore does not have any stockholders. The "profit" that OCLC makes cannot be distributed to individual owners or stockholders, of whom OCLC has none. All the "profit" must stay with OCLC for doing information-related research, developing new information services, and/or enhancing existing services to benefit libraries around the world that OCLC serves.

Although OCLC provides information services to end users such as researchers and teaching professors, students, librarians, and the general public, it always provides such information services through libraries in order to support the important role libraries play in information services. In other words, OCLC does not "sell" information services directly to end users.

OCLC CHARTER

The purpose or purposes for which this corporation was formed are to establish, maintain, and operate a computerized library network and to promote the evolution of library use, of libraries themselves, and of librarianship, and to provide processes and products for the benefit of library users and libraries, including such objectives as increasing availability of library resources to individual library patrons and reducing the rate of rise of library per-unit costs, all for the fundamental public purpose of furthering ease of access to and use of the ever-expanding body of worldwide scientific, literary, and educational knowledge and information.

OCLC VISION STATEMENT

OCLC will be the leading global library cooperative, helping libraries serve people by providing economical access to knowledge through innovation and collaboration.

MEMBERSHIP

OCLC is a membership organization. Any library in the world may become an OCLC user by using any or all information services provided by OCLC.

OCLC is an organization of libraries, governed by libraries, and for the benefit of libraries. Therefore, libraries only can register as OCLC users. Individual persons or commercial companies may not register as OCLC users although corporate libraries (for example, the IBM Library) are welcome to become OCLC users. OCLC user libraries include college and university libraries, school libraries, libraries of research institutions, national libraries, state or provincial libraries, public libraries, special libraries (such as hospital libraries, music libraries, theological libraries), corporate libraries (namely, libraries of commercial companies, for example, the IBM Library), libraries of government agencies, and library consortia. In short, all kinds and all sizes of libraries around the world can be OCLC users. In fact, OCLC membership goes beyond libraries to also include museums.

Any library that uses any of the OCLC information services is an OCLC user. However, OCLC has three categories of users as defined by OCLC Members Council and ratified by OCLC Board of Trustees. The three categories of OCLC users are (1) Governing Members, (2) Members, and (3) Participants.

OCLC Governing Members

Governing Members are the libraries that believe in library cooperation and make a commitment to do at least the following three things: (1) use OCLC online cataloging to catalog all newly acquired titles added to the library collection, (2) attach the holding symbol in WorldCat (the OCLC Online Union Catalog) for the purpose of resource sharing, and (3) contribute original records in WorldCat to facilitate cooperative cataloging among libraries worldwide. Library cooperation is the foundation of OCLC, and contribution of intellectual property is the key to OCLC membership. Every library contributes what it can and benefits from contributions of other libraries. The Governing Members are encouraged but not required to use OCLC ILL, QuestionPoint (electronic virtual reference service), OCLC FirstSearch (online access to databases), OCLC ECO (eJournals), OCLC netLibrary (eBooks), and/or other OCLC services.

OCLC Members

Members are the OCLC user libraries that do not wish to make a commitment to catalog on OCLC all newly acquired titles added to the library's collection but wish to catalog some of their new acquisitions on OCLC and/or to become an OCLC Interlibrary Loan (ILL) Supplier, lending library materials to other OCLC users through the OCLC ILL Service. Libraries around the world are welcome to use OCLC ILL to borrow materials from other libraries. However, OCLC ILL borrowers are not OCLC Members if they do not do any cataloging on OCLC or lend library materials to other libraries through the OCLC ILL Service. OCLC Members may also use other OCLC services as they may wish.

OCLC Participants

Participants are the libraries that choose not to contribute through cooperative cataloging and/or ILL but wish to use other OCLC services. Libraries that use only FirstSearch, ECO, and/or netLibrary eBooks are Participants.

In short, libraries that use OCLC information services can choose to be an OCLC Governing Member, OCLC Member, or OCLC Participant. All of these libraries are OCLC users and may use any or all OCLC services as they wish. However, OCLC Governing Members in many instances pay a lower unit fee than do OCLC Members or OCLC Participants. Besides, only OCLC Governing Members receive credit for original (new) records entered into WorldCat online.

Only OCLC Governing Members can vote to elect Governing Members or Members to serve on the OCLC Members Council. Members can be elected by OCLC Governing Members to serve on the OCLC Members Council, but they cannot vote in the election. Participants cannot vote in the election, nor can they be elected to serve on the OCLC Members Council.

OCLC Members Council

The OCLC Members Council is an OCLC governance body composed of sixty-six delegates elected by OCLC Governing Members from OCLC Governing Members and OCLC Members all over the world. The body meets three times a year in October, February, and May to discuss issues that libraries face and to advise OCLC management. Of the sixty-six delegates, three are elected from Asia and the Pacific region. The number of delegates from each region is based on the percentage of the OCLC information usage from that region. There is a formula developed by the OCLC Members Council for calculation of the representation by region. In the election conducted in April 2004, the following OCLC Members Council delegates and alternates from Asia and the Pacific region were elected to serve a three-year term, October 2004 through May 2007.

Representing Asia and the Pacific:
  Delegate: Dr. Anthony W. Ferguson, University Librarian, The University of Hong Kong Libraries, Hong Kong

Alternate: Ms. Janine Schmidt, University Librarian, The University
   of Queensland, Australia
Representing Japan:
   Delegate: Mr. Yoshiro Kato, Chief Executive, Keio University Library
   (Mita Media Center), Tokyo, Japan
   Alternate: Mr. Makoto Nakamoto, Head, Department of Administra-
   tive Services, Waseda University Library, Tokyo, Japan
Representing Singapore:
   Delegate: Miss Ngian Lek Choh, Deputy Chief Executive, National
   Library Board and Director, Library Support Services, Singapore
   Alternate: Mr. Beh Chew Leng, Chief Executive Officer, eLPEDIA,
   Singapore.

WORLDCAT

WorldCat (the OCLC Online Union Catalog) is the largest database of
bibliographic records and holding information in the world. It is a
global union catalog of about 10,000 libraries. You can find in WorldCat
the bibliographic records of the entire library collections of Harvard
University, Massachusetts Institute of Technology (MIT), University of
California, Berkeley (UC Berkeley), Australian National University, Chinese University of Hong Kong, Hong Kong University of Science &
Technology, The University of Hong Kong, University of Queensland,
Waseda University, all the libraries in Singapore, and thousands of other
large and small libraries around the world. Libraries can access World-
Cat through one of the following three interfaces: (1) FirstSearch, a Web-
based user interface for reference librarians, ILL (Inter-Library Lending)
librarians, and/or library end users including researchers, teaching
professors, students, and the general public; (2) Connexion, another
Web-based user interface for use by acquisitions and/or cataloging librarians for finding and downloading bibliographic records in MARC21
format or Dublin Core; and (3) Z39.50 telecommunications protocol. In
this instance, libraries use the interface of their own library's integrated
systems.

   Regardless of which user interface you may prefer, you need *not* purchase any hardware or software from OCLC in order to access WorldCat.
If you access WorldCat via the FirstSearch or Connexion user interfaces,
the equipment that you need to have is a PC (personal computer), which
shall be loaded with a current version of Web browser, such as Microsoft
Internet Explorer or Netscape. In addition, the PC must be connected to
the Internet. Naturally, your library will need to subscribe to OCLC online

information services and obtain a set (sets) of authorization/passwords from OCLC.

Libraries use WorldCat (the OCLC Online Union Catalog), the largest database of its kind in the world, for the following acquisitions, ILL, and/or reference purposes:

1. To use key-word search or search by Dewey Class Number to compile a comprehensive bibliography on the subject
2. To verify bibliographic and holding information for ILL (Inter-Library Lending) purposes
3. To use WorldCat as a portal to access selected full-text information on the Web, and/or to access full-text eBooks and/or eJournals that your library subscribes to through OCLC
4. To find and download catalog records in MARC21 or Dublin Core format to build the library's OPAC (Online Public Access Catalog) and union catalog of a group of libraries.

WorldCat and Cooperative Cataloging Service

Cooperative Cataloging Service is OCLC's flagship service. No one in the world offers a larger and better online cataloging service than the one offered by OCLC. In order to access WorldCat for cataloging service, libraries shall use Connexion interface, which is a Web-basd interface. Again, users need *not* purchase any hardware or software from OCLC in order to access WorldCat for cataloging purposes.

WorldCat has more than 56 million unique bibliographic records in MARC 21 format and Dublin Core. Records in WorldCat come from the following sources:

1. U.S. Library of Congress—an OCLC cataloging user
2. U.S. National Library of Medicine—an OCLC cataloging user
3. U.S. National Agriculture Library—an OCLC cataloging user
4. U.S. Government Printing Office—an OCLC cataloging user
5. British Library
6. National Library of Canada
7. More than 10,000 OCLC cataloging libraries around the world, including college and university libraries, research libraries, national libraries, public libraries, school libraries, and corporate libraries, etc.

The Library of Congress is an OCLC cataloging user and *all* of the roughly seven million LC-MARC records are in WorldCat. LC-MARC records contain an OCLC Control Number, as these LC-MARC records come from WorldCat. Compared with more than fifty-six million unique catalog records in WorldCat, LC-MARC records represent less than 13 percent of the total number of MARC records in WorldCat. In other words, more than 87 percent of the MARC records in WorldCat are not available through LC-MARC records.

OCLC upgrades LC-CIP (Library of Congress Cataloging in Publication) records. As a result, when you access WorldCat for cataloging, you access LC records not only much faster than through any other channel, but also have more complete LC records than you can find through other channels.

OCLC cataloging service is not merely a matter of locating and downloading MARC records. More importantly, it is a matter of the internationalization of your library's bibliographic information. In addition, it is also a matter of making your library's bibliographic information available to end users on the World Wide Web through Google and Yahoo!

By accessing WorldCat for cataloging of your library's collection, your library's bibliographic information becomes part of the global union catalog in WorldCat, and your library may participate in the global resource sharing. There is no other system in the world that can let you achieve this goal, regardless how much you are willing to pay.

Accessing WorldCat, the largest bibliographic database in the world for cataloging, libraries find the highest hit rate, and thus greatly increase a cataloger's productivity. Human resources are valuable and costly. The increased staff productivity as a result of accessing WorldCat exceeds the cost of using WorldCat.

Library may use WorldCat to catalog titles in roman alphabet (English, French, German, Spanish, etc.) and/or in Chinese, Japanese, and Korea (CJK) characters as well as in Arabic script.

BENEFITS OF OCLC CATALOGING SERVICE TO LIBRARIES

1. Libraries using OCLC Cataloging Service conform to the international cataloging standards such as MARC21, AACR2, LC or DDC Classifications (although OCLC users may use other classifications), LC Authority File, and LC Subject Headings or MeSH (Medical Subject Headings), etc.
2. Access to WorldCat, the largest database of catalog records in the world, results in the highest hit rate, and thus reduces the cost and time of doing original cataloging. It takes a trained cataloger one to three hours to do original cataloging of one title. By comparison, it takes a catalog staff about ten to twenty minutes to catalog a title on OCLC.

3. Access to WorldCat, the largest database of its kind in the world, will satisfy all of your cataloging needs, and, as a result, your library's cataloging staff need *not* spend valuable time searching many small databases just to find a MARC record. A large portion of cataloging cost is the cost of the staff. Staff is a very valuable asset and therefore, staff time shall not be wasted for looking for a record in many small databases. Library staff shall carry on duties that are more challenging and valuable than looking for a record in many small databases each and every day!

4. Since the Library of Congress is an OCLC cataloging user, libraries around the world can access LC-MARC records much faster through WorldCat than through other channels.

5. Since OCLC upgrades LC-CIP (Library of Congress Cataloging in Publication) records, libraries around the world can access higher-quality LC-MARC records through WorldCat than through other channels.

6. OCLC Cataloging Service users can also access LC Authority File.

7. LC Authority File resides on the OCLC server.

8. OCLC does authority update to records in WorldCat from time to time. This benefits OCLC cataloging users when they download records from WorldCat.

9. OCLC Cataloging Service allows users to do authority control online while doing cataloging.

To use OCLC Cataloging Service, libraries need not purchase any hardware or software from OCLC. Libraries do, however, need to have a personal computer (PC), loaded with the current version of Web browser such as Microsoft Internet Explorer or Netscape, and connected to the Internet.

In order to download records into your library's integrated system in real time, you need to work with the developer of your library's integrated system. OCLC cannot and does not work on your library's integrated system.

Libraries may use their library's integrated system to access WorldCat via Z39.50 telecommunications protocol for cataloging or reference purposes.

WorldCat and Sharing of Resources among Libraries

WorldCat contains more than 56 million unique records, covering every academic discipline and all types of materials, including books, serials, AV materials, electronic information, maps, manuscripts, music, etc., in more than four hundred languages. In addition, WorldCat has about one billion holding library symbols attached to these 56 million records, each symbol indicating a location (namely, a library) where the library information can be found. This unique global union catalog is a powerful tool for locating information sought by professors, researchers, students, and librarians. Without this holding library information, sharing of library information resource is practically impossible.

OCLC not only collects and displays this mighty holding information of nearly ten thousand libraries around the world but also develops and maintains a Web-based ILL (Interlibrary Loan) system that processes 25,000 to 30,000 ILL requests a day, or nearly ten million ILL requests a year among OCLC ILL users.

Libraries looking for information not in their own holdings can find the location of that information in WorldCat and send an ILL request to the holding library, which will send the information to the borrowing library via FAX, mail, or e-mail.. There are two types of ILL materials, namely, the returnable and nonreturnable. If the material on loan is a book, that book shall be returned to the lending library within a specified time period. This is a returnable ILL material. However, if the lending material is a photocopy of an article from a journal, that material need not be returned to the lending library, and thus is nonreturnable material. There are also document supply agencies providing document supply services to libraries around the world through OCLC ILL Service.

LIBRARY CONNECTION THROUGH GOOGLE AND YAHOO!

Today's information seekers are going to the World Wide Web for information, bypassing libraries because they think that the Web is the library of the twenty-first century. The Web is undeniably a valuable resource, but it contains along with some gold mines a lot of rubbish. Unfortunately, many information seekers cannot differentiate gold mines from rubbish found on the Web and do not know that libraries contain information only selected by professionals. OCLC has a program called "Find in a Library," in cooperation with Google and Yahoo! (the largest worldwide search engines on the Web), to take libraries to the World Wide Web, close to today's information users. When users search information on the Web using Google or Yahoo! they will find the information in a specific library among the search results. This service is free of charge to libraries that meet the following two criteria:

1. The library's entire holding information is in Worldcat.
2. The library subscribes to WorldCat through OCLC FirstSearch Service.

When information users search on the Web through Google or Yahoo! the library's holdings related to the search term are displayed with "Find in a Library" at the beginning of the entry among the search results. This indicates to the user that this information can be found in a library and is selected by a professional librarian. Users can limit the search result by entering "Find in a Library," a country's name, a city's name, and/or a postal code (e.g., zip code in the USA) in order to find information held in a library or libraries in the user's vicinity. By doing so, OCLC is bringing users back to the library through World Wide Web and Google and Yahoo! Today only OCLC offers this powerful service to OCLC's member libraries.

ONLINE ACCESS TO E-BOOKS, E-JOURNALS, AND DATABASES

netLibrary eBooks

On January 25, 2002, OCLC acquired netLibrary, which was the e-book pioneer and is the leading e-book provider in the world. Now netLibrary has more than 90,000 e-book titles and about ten thousand libraries around the world are netLibrary eBook users, many of whom are in Asia and the Pacific region. In fact, there are netLibrary eBook consortia formed in Australia, China (Mainland), Korea, Taiwan, and Thailand.

To access netLibrary eBooks, libraries need to have a PC, loaded with the current version of a Web browser and Acrobat reader, and be connected to the Internet.

Electronic Collections Online (ECO)—Online Access to E-Journals

OCLC FirstSearch Electronic Collections Online (ECO) provides online access to five thousand electronic journals by more than sixty publishers. It is integrated with OCLC FirstSearch Service, which provides access to more than seventy databases. When users find an index to an article in one of the OCLC FirstSearch Service databases, and when the electronic journal referred to is available on the Electronic Collections Online, the OCLC FirstSearch Service search-result screen will provide a full-text icon linking to the full text on the Electronic Collections Online.

The Electronic Collections Online archives the electronic journals, and provides users with archival access to the volumes of the electronic journals the library subscribed to previously, even after the library has cancelled current subscription to the journal, for as long as the library maintains a valid ECO account.

Archival Access to E-Journals after Subscriptions Cancelled

Archival access is different from access to back issues. Access to back issues means that when your library subscribes to the current year of an e-journal, you may also access that same e-journal's content of the previous years. All publishers allow libraries to access back issues if the library subscribes to the current issue. Some publishers charge an additional fee for accessing the back issues, and others do not. In other words, every publisher allows libraries to access back issues. However, access to back issues is not the same thing as archival access.

If and when your library cancels the current subscription to an electronic journal, the publisher or the e-journal provider cancels your library's access account. As a result, your library can no longer access the issues of the e-journal that your library subscribed to previously. *Only* OCLC allows a library to continue to access the issues that that library subscribed to previously through the Electronic Collections Online (ECO) even after the library has cancelled the current subscription to the e-journal. Naturally, that library would have to maintain an active ECO by subscribing to at least one electronic journal.

Pay per Article to View Electronic Journals

If a library does not wish to pay annual subscription fees to view certain electronic journals for twelve months, it can pay for each article in the journal it needs to view through OCLC ECO and does not have to subscribe to that e-journal. The price of each article is set by the publisher and is displayed online on the OCLC ECO. The per-article price varies from article to article and from publisher to publisher, but it costs between $10 to $35 per article. OCLC users will know how much the article costs before viewing it. Naturally, there is no additional charge to print out, e-mail, and download the article.

OCLC FIRSTSEARCH–ACCESS TO DATABASES AND DOCUMENT SUPPLY SERVICE

The most popular end user online service in university libraries is still OCLC's FirstSearch.—*Library Journal* (October 1, 2001) 41

OCLC FirstSearch Service is the most popular online information retrieval service in the world, designed for use by end users, including professors, researchers, students, librarians, and the public. It provides online access to electronic information of more than seventy databases and about ten thousand full-text journals through Web interface via the Internet. It is available nearly twenty-four hours a day, seven days a week. More than twenty thousand libraries in more than sixty countries use OCLC FirstSearch Service.

OCLC FirstSearch Service offers several pricing options. Libraries that use OCLC FirstSearch Service can pay a fee per search (called Block Purchase, one block = 500 searches). Libraries may purchase and prepay for any number of blocks of searches (namely, 500 searches, 1,000 searches, or 1,500 searches, and so on, in 500-search increments). Libraries may also choose to prepay a fixed price per year (called Annual Subscription) for an unlimited number of searches in twelve months, or use a combination of these two pricing options.

The rule of thumb is that it is cheaper for the user to pay one annual subscription fee to access the databases that are heavily used and pay on a per-search basis (block purchase) to access databases that are not heavily used.

If the user library prefers the per-search (Block Purchase) option, the minimum cost of becoming an OCLC FirstSearch user is $550 to purchase one block of 500 searches, averaging $1.10 per search. Naturally, expanding the library's electronic information services would cost more, and the base price is subject to change.

The OCLC FirstSearch Service charge is *not* based on connect time, and there is no additional charge for displaying, printing, downloading, or e-mailing the search result. The cost of using OCLC FirstSearch Service is usually cheaper than purchasing the same databases on CD ROM.

In addition to indexes and abstracts, OCLC FirstSearch Service contains ten thousand full-text and full-image journals online, including ten million full-text and full-image articles. If full-text articles are not available online, OCLC FirstSearch Service users may request full-text articles through OCLC ILL Service. OCLC FirstSearch Service displays the name of libraries around the world that hold a copy of the material in their library collection.

The British Library Document Supply Centre (BLDSC), Canada Institute of Science and Technology Information (CISTI), the Library of Congress (LC), and many other institutions provide document supply service through OCLC ILL.

To use OCLC FirstSearch Service, users must have a PC (personal computer) connected to the Internet and loaded with the current version of a Web browser such as Microsoft Internet Explorer or Netscape. Users do not need to use any OCLC software.

QUESTIONPOINT (A COOPERATIVE ELECTRONIC VIRTUAL REFERENCE SERVICE)

QuestionPoint is a Web-based cooperative electronic virtual reference service that OCLC introduced in June 2002. It is available 24/7. Interested libraries work together to create and maintain a knowledge base to pool and share the knowledge of reference librarians around the world.

OCLC AND LIBRARIES IN ASIA AND THE PACIFIC REGION

In 1967, OCLC was founded as the Ohio College Library Center, established to serve fifty-four college and universities libraries in the state of Ohio. In other words, the acronym OCLC stands for original name, Ohio College Library Center. Fourteen years later, in 1981, OCLC adopted the current name, OCLC Online Computer Library Center, Inc., and became international by extending its information services to Europe. Five years later, on August 1, 1986, the office of OCLC Asia Pacific was set up to extend OCLC's information services beyond Hawaii to Asia and the Pacific region. At that time, OCLC had no users in Asia or the Pacific region, and the Internet was not yet born. Today, more than 3,600 libraries in eighteen countries and territories in Asia and the Pacific region have online access to OCLC. In fact, a majority (namely, more than 50 percent) of the academic libraries in the following eight countries and territories in Asia and the Pacific region are OCLC users: Australia, China (the Mainland), Hong Kong, Japan, Korea, New Zealand, Singapore, and Taiwan.

The mission of the OCLC Asia Pacific is to extend OCLC's information services to libraries in Asia and the Pacific region, to support library cooperation at the country/regional level, and to link these country/regional cooperatives into OCLC global cooperative. OCLC Asia Pacific covers the territory west of Hawaii, ranging from China, Korea, and Japan in the north to Australia and New Zealand in the south, and India and Pakistan to the west. OCLC's Asia Pacific user base has increased in the past eighteen years from zero in 1986 to about 3,600 libraries in 2004. They are primarily institutions of higher education.

In addition to working with individual institutions, OCLC Asia Pacific has also established a close working relationship with the following key consortia throughout the region.

Australia
- CAUL (Council of Australian University Librarians)
  CAUL sponsored OCLC Asia Pacific's introduction of OCLC First-Search Service to all of the thirty-eight universities in Australia in 1994, and through CEIRC (CAUL Electronic Information Resource Committee), a netLibrary eBooks purchase consortium was formed in 2004.

China (the Mainland)
- CALIS (China Academic Library and Information System)

Through the sponsorship of Tsinghua University Library, Beijing, CALIS has provided online access to OCLC FirstSearch Service to a thousand universities in China since 1999.

- CLC (China Library Consortium)
Through the sponsorship of Peking University Library, Beijing, CLC (China Library Consortium) introduced in 2004 netLibrary eBooks to more than sixty libraries in China.

Hong Kong
- JULAC (Joint University Librarians Advisory Committee)
JULAC is a forum to discuss, coordinate, and collaborate on library information resources and services among the libraries of the eight tertiary education institutions funded by the University Grants Committee (UGC) of the Hong Kong Special Administrative Region Government, People's Republic of China.
JULAC has developed a Chinese Authority File called HKCAN (Hong Kong Chinese Authority—Names), and OCLC Asia Pacific is in negotiation with JULAC to make HKCAN available to all interested libraries worldwide through OCLC. The service is expected to become available through OCLC in 2005–2006.

Japan
- NII (National Institute of Informatics)
NII is a government agency of Japan, operating a library information system called NACSIS. NII and OCLC entered into an ILL agreement in 2002 to establish an ISO system-to-system link of Interlibrary Loan service linking research libraries in Japan and the United States.

Korea
- KERIS (Korea Education and Research Information Service)
KERIS is a government agency of the Ministry of Education, the Republic of Korea, and has provided online access to OCLC for three hundred university and research libraries since 1997, including cataloging, FirstSearch, ECO electronic journals, and netLibrary eBooks.

New Zealand
- CONZUL (Council of New Zealand University Librarians)
CONZUL sponsored OCLC Asia Pacific's introduction of OCLC FirstSearch Service to all of the eight universities in New Zealand in 1994.

Singapore
- SILAS (Singapore Integrated Library Automation Service)
SILAS is a Singapore government agency providing information services to all libraries in Singapore. OCLC loaded 1.4 million records from the Singapore National Union Catalog into WorldCat in 2002, and all of the 101 libraries in Singapore became OCLC Governing Members in the same year, cataloging all of their newly acquired titles

on OCLC. This program has internationalized the National Union Catalog of Singapore.

Taiwan

- CONCERT (Consortium on Core Electronic Resources in Taiwan)
  Managed by STIC (Science and Technology Information Center, National Science Council, Taiwan), CONSORT has provided online access to OCLC FirstSearch Service to more than 100 universities in Taiwan since 1999.
- TEBNET (Taiwan eBook Network)
  Through the sponsorship of Feng Chia University, Taichung, Taiwan, TEBNET has provided online access to netLibrary eBooks to about forty institutions in Taiwan since 2002.

Thailand

- Thai University eBook Net and UNINET
  Two netLibrary eBook purchase consortia, Thai University eBook Net and UNINET, are being formed and are expect to materialize by the end of 2004.

Input of Asian Information into WorldCat

Aside from delivering information in the English language to libraries in Asia in support of education and research, OCLC Asia Pacific has also facilitated input of Asian information into WorldCat (the OCLC Online Union Catalog) for sharing among libraries worldwide. Waseda University, Tokyo, Japan has batchloaded 780,000 Japanese records and 1.4 million holdings since 1995 and is now sending about three thousand Japanese records to OCLC monthly. As a result, the entire collection of Waseda University Library is in WorldCat, and Waseda University Library functions as a document supply center of Japanese materials to research libraries worldwide, particularly those in the United States, through OCLC ILL Service.

Libraries in Hong Kong, particularly Chinese University of Hong Kong, Hong Kong Baptist University, Hong Kong Polytechnic University, Hong Kong University of Science & Technology, and the University of Hong Kong are major online contributors of Chinese records for materials published in China (the Mainland), Hong Kong, and Taiwan.

CONCLUSION

Although information knows no national boundaries, nationalism is real, and it is both an asset and a barrier as far as sharing of information on a global scale is concerned. In addition to nationalism, the following are also determinant factors of sharing of information, namely: economy, infrastructure, culture that detects the value of information, and language.

The demand for OCLC's information services in Asia and the Pacific will continue to grow because of the increasing popularity and bandwidth of the Internet, increasing demands of end users to access electronic information in full-image or full-text, and the projected economic growth in this region in the years to come. In order to facilitate this increasing demand, OCLC has extended the online system operation around the clock, developed user interface in Chinese, Japanese, and Korean (CJK) characters in FirstSearch and netLibrary, and Chinese and Korean interface in QuestionPoint. In addition, OCLC has a plan to support UNICODE and incorporate Asian contents in OCLC services in 2005, particularly in FirstSearch and netLibrary.

Chapter 3

The Impact of Information Technology on University Libraries in India

T. A. V. Murthy and V. S. Cholin

ABSTRACT

Information technology (IT) has revolutionized the information handling activities in research and academic libraries in India. The university libraries, as centers of information services, have largely been affected by the rapid changes in IT. Indian universities constitute one of the largest higher education systems in the world, comprising more than 310 universities/institutions, 14,000 affiliated colleges, and ten million students, with half a million teachers across the country. The university libraries in India are at some stage of development in the application of information technology tools in their day- to-day activities. All academic libraries now virtually depend on the IT systems for their basic operations such as acquisitions, cataloguing, circulation, serials control, and other functions.

The library gives access to vast information sources, not only the items held by or owned by the library but also, by providing access to them, to remote information sources. This chapter is an attempt to give an overview of university libraries in India, with the main emphasis on the effort initiated by the University Grants Commission (UGC) through the Information and Library Network Centre (INFLIBNET) (an inter-university center of the UGC) located at Ahemdabad in the western part of India to support the university libraries in automation and networking. The article also highlights the recent initiative of UGC-Infonet, which is likely to make remarkable changes in collection development at universities and also in providing effective and easy access to information resources through the UGC-Infonet E-Journals Consortium.

INTRODUCTION

Libraries play a pivotal role in ensuring the success of upper-level research. The important activities of university libraries include collection development, reference service, document delivery, user education, and access to electronic resources. With the development of Information Communication Technologies (ICTs), university libraries are expected to provide cost-effective and reliable access to information using state-of-the art information technology tools. Information technology has revolutionized the information handling activities in the academic libraries during the past few years. The developments in the IT sector have proven the death of distance and the death of time. University libraries are fertile areas for the introduction of IT for providing and making accessible to the user community the best possible information from anywhere at any time and from any sources. Networking will be the essential partner in this exercise, as it facilitates access to vast information services. Networks have potential to improve library services in several ways.

The continuous improvement in networking technology helps libraries to reduce the cost of supplying information, thus creating new opportunities for libraries to serve their end users. In recent years libraries worldwide have been affected by an uncertain financial environment in which resource buying has been restricted, causing them to look at ways of extending their purchasing capabilities to compensate for reduced budgets. The situation is one of "United we stand, divided we fall." Library and information centers are increasingly being called upon to provide more relevant, up-to-date, and timely information to a wide range of users. To satisfy the varied needs they require availability and accessibility to a variety of information resources and formats. The libraries in most developing countries suffer from inadequate funding or stringent budget cuts. This has affected the level of services offered to users, both in terms of quality of collections and the degree of staff support provided. In the present circumstances only a few libraries can budget a wide range of information resources. Universities/institutions have realized that the concept of organizational self sufficiency has been replaced with collaborative survival, as more and more new information resources are generated while funds decrease. The situation calls for changes in the collection development approach and for avoiding duplication of information resources among the libraries in the country; it is wiser and more cost-effective to share resources via cooperative purchases, in a consortium.

APPLICATION OF INFORMATION TECHNOLOGY
IN ACADEMIC LIBRARIES.

Perhaps the greatest example of vision and focus in automation over the past quarter-century was the development and rise of the MARC format. The developers of MARC recognized the need to communicate bibliographic information in a standard format. The format is highly flexible and amenable to change. The MARC format was created in a world where librarians had to generate cataloging information and the standards for catalog records. Automation of library card catalogs provides a finding tool for the library collections. The books, journals, films, and other materials located through the catalog still mostly reside in their original form, with no direct connection to the automated finding tool. Most of the early development in electronic publishing was also aimed at identifying information sources. Publishers of indexing and abstracting serials were the first to provide their resources in electronic form. In the 1970s and 1980s, indexing and abstracting databases were predominant in the online database world. Bibliographic databases are still the most widely used type of electronic resource in libraries, even now.

The Library of Congress's initiatives with regard to data formats facilitated the creation of OCLC and other bibliographic utilities. OCLC's initial objective was to provide libraries with shared, and hence affordable, access to automation for cataloging and the production of catalog cards. While this is still a significant activity, today over 75 percent of OCLC's full cataloging participants download cataloging records into their local library systems rather than obtaining cards, and nearly 80 percent use the interlibrary loan subsystem for resource sharing—currently at the rate of one million interlibrary loans every two months.

The electronic resources available in libraries today are an outcome of advances in computer technologies, with powerful computers and information storage and delivery mechanisms, such as CD-ROMs and user-friendly interfaces. In most of the academic libraries in Western countries, Online Public Access Catalogues (OPACs) have almost replaced card catalogs, offering enhanced search capabilities for accessing the local collection and often expanding coverage to include the holdings of other area or regional libraries. Many libraries in India also provide a Web interface to their library and information system. The library or information system with a Web interface often includes direct links to electronic journals, books, and Internet resources. Modernization of university libraries in Indian universities started in the early 1990s, but progress in the beginning was very slow, if steady. In the majority of the university libraries, computers were most reluctantly accepted because of a lack of competence among professionals. With the advent of INFLIBNET of

UGC, the All India Council for Technical Education (AICTE), the National Information Systems for Science and Technology (NISSAT, now renamed the National Institute of Science Communication and Information Resources [NISCAIR]), and other organizations of national stature with a good blend of traditional IT skills gave university libraries a necessary facelift in terms of modernization. Now, with the use of PCs and CD-ROMs for developing local databases for literature searches, Internet connectivity is quite common in almost all the universities, making the automation activity smooth.

Most of the libraries have traditionally tried to own resources as much as possible, because owning an item spares the patron delays for borrowing or purchasing on demand. However, the increased cost of maintaining a collection of primary sources and the increased demand for information has resulted in a shift in emphasis from ownership to access. In the present situation, academic libraries in India have been largely affected by financial constraints that restrict resource acquisition. Most university libraries are ill equipped to satisfy user needs within their resources, given the exponential increase in information. Price escalation and high foreign exchange rates force libraries to discontinue subscriptions to many publications. At the same time there is a deliberate, seemingly inevitable duplication of costly library holdings in the absence of convenient sharing mechanisms. Scholars in remote areas feel mentally isolated. It is also impossible to fund all libraries to make them self-sufficient to meet resource requirements. And those are just some of the challenges facing the academic libraries.

UNIVERSITY LIBRARIES IN INDIA

There are more than 310 university-level institutions in India (including forty-two deemed universities). Of these, 161 are traditional universities (including institutions for specialized studies in academic disciplines), while the others are professional/technical institutions. Of the latter, thirty-four provide education in agriculture (including forestry, dairy, fisheries, and veterinary science), eighteen in medicine, and twenty-five in engineering and technology; there are ten so-called open universities. There are specialized institutions also, which include seven Sanskrit universities, five women's universities, seven universities for regional languages, four for law, and one each for population sciences, music and fine arts, statistics, and journalism (*Universities Handbook* 2002).

There are around 150 universities eligible for grants from UGC for automation and networking. The present discussion deals with those universities. Under UGC rule 12 (ccc), some universities—for example, agriculture universities and medical universities—are not eligible to receive grants and are not covered here.

Indian universities have in fact been using IT tools for providing effective information services for some time. Most university libraries in India are at some stage of development in the application of IT tools in their day-to-day activities. All academic libraries virtually depend on the IT systems for their basic operations such as acquisitions, cataloguing, circulation, serials control, and so on. The library provides access to a vast information sources, not only the items held by or owned by the library but also access to remote information sources. Of course, they must also handle the resultant requirements to authenticate and authorize users. These are the key challenges for the modern academic librarian (Arant and Payne 2001).

CREATION OF INFLIBNET—A BEGINNING

The Information and Library Network Centre (INFLIBNET) is the culmination of a yearlong effort of an interagency working group. This group, of which T. A. V. Murthy is a member, consists of experts from library science, computer science, and communication who have worked diligently to prepare the sketch of INFLIBNET. A blue book was prepared and submitted to UGC in 1988 with recommendations, many of which are still valid after more than fifteen years. All these recommendations deal with one major objective, which is resource sharing to facilitate optimum utilization of resources by various methods. INFLIBNET started as a project under the Inter-University Centre for Astronomy and Astrophysics (IUCAA) in 1991, with its headquarters at Ahmedabad, and became the independent Inter-University Centre (IUC) of UGC in 1996 with its own governing council and governing board as advisory bodies and its own director, the executive head who manages the center's activities.

Over the years, the program has progressed steadily, and since May 1996 it has been an independent, autonomous Inter-University Centre under UGC to coordinate and implement a nationwide high-speed data network using state-of-the-art technologies for connecting all the university libraries in the country. INFLIBNET has set out to be a major player in promoting scholarly communication among academicians and researchers in India.

Objectives

Objectives and functions of INFLIBNET as envisaged in the Memorandum of Association are:

- To promote and establish communication facilities to improve capability in information transfer and access, that provide support to scholarship, learning, research and academic pursuit through cooperation and involvement of agencies concerned
- To establish INFORMATION AND LIBRARY NETWORK "INFLIBNET"—a computer communication network for linking libraries and information centres in universities, deemed to be universities, colleges, UGC information centers, institutions of national importance, and R&D institutions, etc., avoiding duplication of efforts.

The Role of INFLIBNET in Automation of University Libraries

- To bring the IT culture to the universities and automate the university libraries
  - INFLIBNET with the support of UGC has spent several crores of rupees by giving the initial grant and subsequent grant for five years to these universities. This helped the libraries substantially to procure the hardware and software for library automation activities.
  - INFLIBNET, during the last ten years, had ups and downs and passed through its teething troubles. However, it is developing and fast expanding in size, resources, and services. Some significant results are visible and noteworthy, as discussed below. This progress has been made possible only by the continuous support of the parent body, UGC, and the participating libraries.
    - Provided financial support to the tune of Rs. 6.5 lakhs each to 142 university libraries for the purpose of automation and networking. Of these, more than 90 percent have become operational and started implementing the recurring grant for the next five years.
    - Provided to core facility grant of Rs. 1 lakh each to sixty-five libraries to establish core facilities and get connected to the network for accessing the information resources.
    - Conducted intensive training courses and workshops for the staff working in these libraries. Apart from this, on-site training has been provided at around three dozen locations.
    - Provided proper guidelines for the creation of quality records as prescribed by the expert committee recommendations, and copies of the guidelines were given to all these libraries.
    - Conducted more than forty INFLIBNET Regional Training programs on Library Automation (IRTPLA) at different states covering more than 800 college librarians.
    - Developed and supplied Software for University Libraries (SOUL) the state-of-the-art library management software, to

more than 115 libraries. Implementing the MARC-21 interface to SOUL software and vice versa.

- Developed national databases of different materials from universities and provide access through The inflibnet website is www.inflibnet.ac.in. There is a quantum jump in the database of books with holdings of more than twelve lakhs of records. User-friendly search engines have been developed to provide access to these databases.
- Created databases of experts and projects and provided access to the data concerning more than 12,000 experts. Under the project of NISSAT INFLIBNET has created more than 20,000 expert profiles in the field of science and technology and made them available on the website http://nissat.inflibnet.ac.in/ and also developed software facilitates online for uploading and editing of profiles.
- Provides occasional technical guidance to all the libraries for implementation of IT.
- Developed Library Management Software called SOUL, which is installed at more than 450 institutions/colleges.
- Provides information services to the research and academic community using the CD-ROM databases, access to OCLC, First Search, STN International, etc.
- Conducts an annual convention to provide a platform for librarians and IT professionals in the form of Convention on Automation of Libraries in Education and Research Institutions (CALIBER), which has become a very important forum for the discussion of modern trends in library science.
- Brings out a series of publications to promote the cause of INFLIBNET.

More importantly INFLIBNET has been able to create an IT-conscious environment in the university libraries. Librarians have now accepted changes and are working to bring about these changes in their libraries.

THE STATUS OF AUTOMATION AND
NETWORKING IN INDIAN UNIVERSITIES

Through the INFLIBNET Centre, the UGC has been a consistent source of funding to Indian universities specifically for automating library activities and networking the resources available in these libraries. The universities funded are expected to report the status of the library automation and networking every year. To date there have been 142 universities di-

Table 3.1.   Status of Computerization

Status of Computer System	Number of Universities
Systems Installed	126
Yet to Inform of the Status	16

rectly benefited by the UGC for library automation and networking. These universities have also been provided with recurring grants for equipment maintenance, the salary of an information scientist (wherever appointed), data entry support for five years, telephone network usage cost, and consumables.

Automation of University Libraries

Under INFLIBNET, the 142 funded universities have been provided with a nonrecurring grant of Rs.6.5 lakhs and a recurring grant for five years to cover the expenses of maintenance, the salary of an information scientist, if any, support for data entry, telephone charges for data access, consumables, and so on.

Some universities have not reported their status, but most of them have installed their computer systems in the libraries.

Status of Database Activity

Based on the guidelines provided by the INFLIBNET Centre, university libraries have been creating records in machine-readable form. Almost all universities have started the database creation in their universities but very few have processed their complete collections, and many other universities are still in the process of cataloging their collections in machine-readable form.

Table 3.2.   University Database Activities

Action	Database Creation
Started	127
Yet to Inform of the Status	16

Database Creation (Range of Records)	Number of Universities
More Than 1 Lakhs	20
50,000–100,000	19
10,000–50,000	28
Less Than 10,000	24
Yet to Inform of the Status	51

Expert Database at INFLIBNET

Around 15,000 expert profiles for the faculty members working in the universities are available in the database form, and many more profiles are yet to be included. The center has taken on a project of creating the expert manpower module in Science and Technology funded by NISSAT, Department of Scientific and Industrial Research (DSIR). Under the project, 16,000 curriculum vitae have been compiled and are available through online access from INFLIBNET website (www.inflibnet.ac.in).

HUMAN RESOURCE DEVELOPMENT

- INFLIBNET to date has conducted twenty four-week training programs for the operational staff working in the universities covering 400 participants from universities receiving the grants
- Conducted seven one-week workshops for executives working in the university libraries covering more than 140 participants
- Conducted thirty-one on-site visits and trained the staff at these libraries
- Conducted more than forty regional training courses covering more than a thousand college librarians on the use of SOUL software developed by INFLIBNET and other automation related topics
- Conducted a number of SOUL orientation training courses at Ahmedabad
- Conducted the SOUL Familiarization Programme at three places—Mumbai (ninety librarians); Kolkatta (170 librarians), and Chennai (125 librarians). Demo CD of SOUL software was distributed during the program
- Six training programs were conducted on "E-resource management using UGC-Infonet."

NETWORKING FACILITIES

Initially all universities were asked to get connectivity using existing data networks (Internet Service Providers, or ISPs) such as the Education and Research Network (ERNET), National Informatics Centre Network (NIC-NET), Videsh Sanchar Nigam Limited (VSNL). Many universities had the dial-up or leased-line connectivity from these Internet service providers. No university could do much with such limited facilities, so the University Grants Commission decided during its Golden Jubilee Celebration year to dedicate the UGC-Infonet Network to the nation to connect the more than 171 universities eligible to receive grants under UGC.

UGC-Infonet Initiative of UGC

The University Grants Commission has launched two ambitious projects called UGC-Infonet and E-Journals Consortium. This will act as boon for the research and academic activity in the country.

UGC-Inofnet Connectivity

The UGC-Inofnet Connectivity initiative was taken to bring about a qualitative change in the academic infrastructure, especially for higher education. Under this initiative UGC is modernizing the university campuses with state-of-the-art campuswide networks and setting up its own nationwide communication network named UGC-Infonet. The INFLIBNET Centre, an autonomous IUC of the UGC, is the coordinating and monitoring agency in the UGC-Infonet Project. It provides liaison between UGC, ERNET, and the universities. INFLIBNET is also responsible for training university library professionals in the use of this network to supply a variety of services to the users. ERNET India, a scientific society under the Ministry of Communications and Information Technology (MCIT), in partnership with the UGC, is setting up UGC-Infonet. UGC-Infonet has been a boon to the higher education systems in several ways:

- UGC-Infonet has become a vehicle for distance learning to facilitate the spread of quality education all over the country.
- UGC-Infonet has been a tool to distribute educational material and journals to the remotest of areas.
- UGC-Infonet has been a resource for researchers and scholars for tapping the most up-to-date information.
- UGC-Infonet has become a medium for collaboration among teachers and students, not only within the country but also all over the world.
- UGC-Infonet has become an Intranet for university automation.
- UGC-Infonet encompasses entire university systems for the most efficient utilization of precious network resources.
- UGC-Infonet has established a channel for Globalization of Education and facilitates the universities in marketing their services and developments.

The status of connectivity under UGC-Infonet is given in appendix A.

UGC-INFONET E-JOURNALS CONSORTIUM

With globalization of education and competitive research, demand for journals has increased over the years. Due to insufficient funds, libraries have been forced to cut subscriptions of journals.

UGC has turned to the Internet to cover the gap between demand and supply by way of e-journals that can be subscribed to online, as most of the journals are available in electronic form. Once the universities are given the connectivity, the access to scholarly journals and databases is made available through UGC-Infonet E-Journals Consortium. The program is wholly funded by UGC and is being executed by the INFLIBNET Centre at Ahmedabad. All universities that come under UGC's purview are members of this program, and it will gradually be extended to colleges as well. Access to various e-journals started January 1, 2004. The program will increase in a very fundamental way the resources available to the universities for research and teaching. It will provide the best current and archival periodical literature, from all over the world, to the university community. The program will go a long way to mitigate the severe shortage of periodicals faced by university libraries for many years, due to the ever-widening gap between the growing demand for literature and the limits of available resources.

The E-Journals Program is the cornerstone of the UGC-INFONET effort, which aims at addressing the teaching, learning, research, connectivity, and governance requirements of the universities. The E-Journals Program demonstrates how communication networks and computers can be used to stretch and leverage available funds in furthering these aims. The program has been made possible by the close and understanding cooperation between the UGC, the Education and Research Network (ERNET), the Inter-University Centres Inter-University Centre for Astronomy and Astrophysics (IUCAA), INFLIBNET and the Consortium for Educational Communication (CEC), and the national and international publishers. A bouquet of e-journals was presented to the nation by His Excellency the President of India on December 28, 2003, during the concluding day of UGC's Golden Jubilee celebrations.

The E-Journals Program aims at covering all fields of learning of relevance to various universities, including:

- Arts, humanities, and social sciences
- Physical and chemical sciences
- Life sciences
- Computer science, mathematics, statistics, etc.

The literature made available includes journals covering research articles, reviews, and abstracting databases. Access is provided to current as well archival literature. The beauty of this initiative is that researchers denied access to important resources five or ten years ago can now access not only to the current literature but also full-text journals back eight to ten years. Portals are also being provided to enable users to navigate easily through all the literature that is made available through this initiative, which serves as one-stop shopping for the research community—a gateway to literature published by all major publishers.

Electronic Resources Subscribed under UGC-Infonet E-Journals Consortium

The UGC-Infonet E-Journals Consortium subscribes to the following resources listed in table 3.1. These resources were made available to fifty universities/institutions per the recommendations of the National Negotiating Committee as of October 2003, with actual subscription starting on January 1, 2004. Fifty more universities are covered under the program with access to similar collections as of September 2004. All electronic resources are accessible from the publisher's website. Many more resources are under consideration and are likely to have been made available as of January 2005.

CONCLUSION

The most important advantage of the information age for libraries may be that the information is not limited to items held by the library, but, rather, the user can access any modern library in the world through the World Wide Web. It is important to use the latest developments in IT, especially the Internet, to improve the quality of collection and information services in the modern library. The university libraries across India have initiated the IT applications in their respective libraries, but the status in respect of automation of their existing collection is not encouraging, as the progress is very slow. This activity is ongoing at the individual universities.

The Indian academic and research community mainly depends on the journals for research work. Even by spending more than 75 percent to 80 percent of the library budget on journal subscriptions, the libraries are not in a position to meet the requirements of its users because of the ever-increasing cost of foreign journals and also the fluctuation in the conversion of Indian rupees against all the major foreign currencies. Hence the universities should be encouraged to access the journals in electronic form on the Web. With the initiative of UGC-Infonet, the research and academic community gets easy access to scholarly journals and databases. The e-subscription initiative under UGC-Infonet is expected to trigger a remarkable increase in sharing of both print and electronic resources among universities across the country. The initiative facilitates not only access to electronic resources subscribed to under the program but also access to resources of other libraries participating in the consortium.

Table 3.3. E-Resources Subscribed under UGC-Infonet E-Journals Consortium

Name of the Publisher	No. of Journals / Database	No. of Universities
1. American Chemical Society	31 journals	50
2. Royal Society of Chemistry	23 journals + 6 databases	50
3. Chemical Abstracts Services (Sci-finder Scholar)	One database	10
Chemical Abstracts—STN service	One database	100
4. Nature Publishing Group	One journal	50
5. Institute of Physics Publishing (IOPP)	36 journals	50
6. Cambridge University Press	72 journals	50
7. Project Muse	222 journals	50
8. Biological Abstracts—BIOSIS	One database	50
9. Encyclopedia Britannica	National site licensing	All Academic Institutions
10. JSTOR—Archival Access	293 journals (Arts & Science I &II, General Science, Languages & Linguistics)	24
11. American Institute of Physics	19 journals including AIP society package-II titles	50
12. American Physical Society	8 Journals	50
13. Science Online	One journal	50
14. Springer and Kluwer Publications	Subscriptions to 200 journals but access to all 1,200 journals in the initial one year	50 + 50 Universities will have trial access
15. Elsevier Science—Life Sciences	34 journals in current trends, opinions, cell press	50
16. Emerald Library Science Collections	28 Journals	30
17. Annual Reviews	29 journals	50
18. Gateway Portals	Cover more than 10,000 journals up to abstract level	56 universities for one year

The library will continue to be a dynamic provider of all types of information. The spectrum of services offered to users will enlarge with IT. With this development, library and information centers are able to get access not only to a wide variety of bibliographic information but also to full-text resources. The INFLIBNET Centre, in its initial phase, has focused on building up the infrastructure for the participating libraries and preparing them to accept the change brought about by IT; it is now focused on providing e-access to resources through the available sources and also working toward providing e-access to Indian content by initiating content creation activities.

NOTES

1. T. A.V. Murthy and V. S. Cholin, "Library Automation" (paper presented at the 1st International Convention on Automation of Libraries in Education and Research Institutions (CALIBER), Ahmedabad, India, February 2003), 1–10.

2. Wendi Arant and Leila Payne, "The Common User Interface in Academic Libraries: Myth or Reality," *Library Hi Tech* 19, no. 1 (2001): 63–76.

3. INFLIBNET Website: www.inflibnet.ac.in (8 June 2004).

4. *Universities Handbook* (New Delhi: Association of Indian Universities, 2002), 1.

Chapter 4

The Digital Age and Information-Poor Societies: The Case of Vietnam

Binh P. Le

ABSTRACT

This chapter examines the impact of information technology (IT) on library services in Vietnam, including the utilization of microcomputers; the development of local and wide-area networks and online electronic databases; and Internet usage. Vietnam possesses an extensive network of libraries. However, with a few exceptions, material resources are inadequate. Moreover, the applications of generally accepted international library standards and practices are also limited. It is suggested that while it is important for Vietnamese libraries to devote their scarce resources to the development of IT, it is equally vital that they pay attention to improvement in resources acquisition and accessibility. Equally important, Vietnamese libraries, especially academic libraries, ought to play a proactive role in articulating information need, curriculum development, and wider and uniform usage of international library standards and practices. as well as cultivating and nurturing a scholarly community.

INTRODUCTION

Vietnam officially entered the digital age on April 14, 1994, when the ".vn" domain was registered with the Internet world. A few months later, the Institute of Information Technology, with the assistance from the Australian National University, established one of the first information networks, called VARANET (Vietnamese Academic, Research and Education Data Communication Network). Recently, according to the World Economic Forum's *Global Information Technology Report, 2003–2004*, Vietnam ranks sixty-eighth among 102 countries on the networking-ready index. The ranking places Vietnam slightly ahead of the Philippines (sixty-ninth) and Indonesia (seventy-third). Technically speaking, this was a remarkable technical transformation because the first and only computer that made its way into the Vietnamese library world was an Olivetti personal computer donated to the National Library of Vietnam by the National Library of Australia in 1986, and the Internet was only a rumor in Vietnam in the early 1990s. Computers and the Internet, far from being prevalent in Vietnam, have gradually penetrated into Vietnamese libraries, and their impact on library services, albeit limited, is also being realized.

This chapter provides an overview of the development of IT in Vietnamese libraries. Its aims are to highlight the impact of and impediments to IT in Vietnam's library services. Whereas this chapter addresses the impact of IT on library services in Vietnam, it should prove useful for those who are interested in the issue of IT and library services in developing countries. They all face similar social, political, economic, and particularly technological challenges.

VIETNAM'S NETWORK OF LIBRARIES

Vietnam established its first library in the eleventh century. However, modern libraries are a relatively recent phenomenon. Indeed, most Vietnamese libraries were established in the second half of the twentieth century, noticeably in the 1980s and 1990s. For example, in 1975, there were eleven public libraries in the south; by 1985 this number had increased to 307 public libraries. Today Vietnam possesses several extensive networks of libraries. Recent data show that there are over twenty thousand libraries of different types, millions of books, and thousands of domestic and foreign periodicals, and more than twenty-three thousand library staff and librarians. For a country that had gone through several decades of war and international isolation, the establishment and maintenance of such networks of libraries are quite extraordinary.

PUBLIC LIBRARIES

By far the largest and best organized library network is the public library system. It consists of the National Library of Vietnam (NLV), sixty-three provincial and municipal libraries, 574 district libraries, and 470 commune libraries/reading rooms. The NLV alone holds over 1,200,000 volumes and 8,000 domestic and foreign periodical titles. The NLV is also the legal depository library of Vietnam, which, by law, requires three copies of every work to be deposited with the NLV.

Over the last decade, the NLV has also established more than 150 mobile libraries to serve rural and ethnic populations residing in the remote regions of the country. To bring library services to these people, mobile library staff in many cases have to utilize whatever means of transportation available, including minibuses, boats, bicycles, motorbikes, and horses, as well as on foot.

In the South, the General Science Library (GSL) of Ho Chi Minh City, formerly known as the National Library of Vietnam, holds nearly one million items, many of which were published between 1954 and 1975. The GSL is a depository library of UNESCO, World Bank, Food and Agricultural Organization, and the International Atom Energy Agency. Most Vietnamese public libraries, except the few large ones, are still underdeveloped.

Interestingly, the extensive network of Vietnam's public libraries was one of the legacies of the Vietnam War. A couple of factors contributed to this phenomenon. First, one of the goals of the Vietnamese Communist Revolution was to eradicate illiteracy. Consequently, books and reading materials were made available to people throughout the country. The other factor was the need to provide information sources, especially ideological materials, to communist cadres. Thus, by the time the war ended, there already existed an "information network" in the country.

ACADEMIC AND SCHOOL LIBRARIES

At last count there were 224 higher education institutions in Vietnam. Each higher education institution has its own library or libraries. As of 2004, there were two hundred academic libraries. The Vietnam National University in Hanoi holds nearly a million items and three thousand domestic and foreign periodicals, but most Vietnamese academic libraries are still far from adequate.

A number of circumstances explain why Vietnamese academic libraries are underdeveloped. First, Vietnamese students, unlike their counterparts in developed countries, especially in North America, are not expected to conduct research. The students' only concern is to pass the examinations, which require the students to do almost nothing beyond relying on the teachers' lectures and textbooks. Second, only a very small number of Vietnamese faculty conduct research and publish, because Vietnamese faculty are not required to publish. The notion of "publish or perish" is not part of the higher education culture in Vietnam. Third, most of the research in Vietnam is conducted by government researchers, who rely mostly on the resources provided by the government's own research institutes. (As we'll see later, this is why Vietnam possesses a very extensive network of specialized libraries.)

In addition to university libraries, there are 1,800 other school libraries, most of which are located in elementary and high schools and vocational schools. In general, most of these libraries are small. In fact, many of these libraries might be considered as "bookcases" and/or "reading rooms."

SPECIALIZED LIBRARIES

Vietnam possesses an extensive network specialized libraries. By and large, each government ministry has its own network of libraries. According to one estimate there are at least three hundred specialized libraries in Vietnam today. The size of the libraries and/or networks varies, depending on the mission and function of each individual ministry—which often reflects the government's policies at the time the library was initiated. In the 1960s and 1970s, the main focus of the government was on the revolution and the struggle against the United States. Consequently, most of the efforts were devoted to the development of social, political, and humanities libraries, which ultimately led to the creation of a network of social sciences libraries. This network of libraries has been placed under the control of the Social Science Commission. At the heart of this system is the National Information Institution of Social Sciences and Humanities (NIISSH) in Hanoi. It is responsible for providing information services, training, guidelines, etc., to all libraries in this system.

When the war ended in 1975, the government's focus shifted toward economic development, especially agricultural development. As a result, most of the efforts were geared toward the development of agricultural and rural libraries. Among the most important entities in this system is the Center for Agriculture and Rural Development (ICARD). ICARD is responsible for providing information on the agricultural sector. Recently, ICARD developed an intranet called AGRUNET, aimed at providing agricultural information sources to its users, especially policy makers. The Central Library of Agriculture, which is under the administration of the Ministry of Agriculture, manages the country's agricultural library network.

In recent years, because of the rapid technological development and Vietnam's desire to join the digital age, more resources have been poured into the development of scientific libraries and information infrastructures. As a result, a very advanced network of scientific and information libraries was established. At the heart of this system is the National Center for Scientific and Technological Information and Documentation (NACESTID). NACESTID, created in 1990, was the result of the merger of the Central Institute for Scientific and Technical Information, founded in 1960, and the Central Library for Science and Technology, founded in 1972. As of 2004, NACESTID consisted of forty-four scientific and technological information and documentation sectoral centers at ministries, state committees, and agencies; fifty-three provincial/municipal scientific and technological information and documentation units; and 250 information and documentation units in universities, high schools, research institutes, publishing establishments, and hospitals. This organization, under the control of the Ministry of Science, Technology and Environment (MOSTE), is responsible for developing as well as disseminating materials relating to economics, science, technology, and the environment to users, especially policy makers.

In addition, the Vietnamese armed forces also possess an extensive network of libraries. In fact, military libraries have been organized at almost every level of the military, from the Ministry of Defense at the top of the military hierarchy down to the company level at the bottom. There exists such an extensive network of military libraries because the Vietnamese armed forces, like the armed forces in other communist countries, place strong emphasis on political and ideological (Marxism/Leninism) education as well as military education among their soldiers.

The military system consists of 1,500 libraries, all of which are under the control of the Central Military Library. According to some estimate, the Central Military Library alone holds a collection of more than three million items and 1,500 periodicals. Many military libraries hold between 200,000 and 300,000 volumes.

Besides these major networks of libraries there are other special libraries in the country. For instance, there are the National Archives in Hanoi, "Luu Tru I" and the National Archives in Ho Chi Minh City, "Luu Tru II," and the various libraries or research entities located within governmental organizations such as the Southeast Asia Institute, the Communist Party's Institute of Party History, and the Foreign Ministry's Institute for International Affairs.

Information Technologies and Library Services

By the late 1980s and 1990s, personal computers (PCs) were in place in some of Vietnam's major libraries and research organizations. The number of PCs available was rather small. During this period, there were only a few library staff members who knew how to use PCs and/or had library automation skills. According to one survey, many librarians expressed "considerable fear about using computers," and "all ten library directors admitted that they had no computer skills of any kind." Actually, PCs were used to perform administrative and related tasks, especially word processing, rather than for information-processing functions.

It was not until 1994 that most of the major Vietnamese libraries and research centers were equipped with PCs. Gradually, more computers were added to libraries. At the same time, the NLV, NACESTID, and other educational and research organizations also offered training programs to help library professionals learn how to use computers as well as computer applications in library services. The early applications of computers in Vietnamese libraries concentrated in the following areas: creating in-house databases (mostly online catalogs), running CD-ROM electronic databases, and forming library networks.

DEVELOPMENT OF LIBRARIES' DATABASES

Following the acquisition of computers, Vietnamese libraries began to build bibliographic databases. The databases that have been created would not have been possible without support from UNESCO. In the middle of the 1980s UNESCO developed a bibliographic utility software program called Computer Documentation System/Integrated Set of Information Systems (CDS/ISIS) and donated it to libraries in the developing countries. Although it is a not user-friendly software program, it has enabled many libraries in the developing countries to build bibliographic databases.

Online catalogs were the first ones to be created. Initially, libraries were able to enter only newly acquired materials. Since then, limited conversion of older materials has been carried out. Most libraries, especially the larger ones that have not yet finished entering all the records of their holdings, continue to use card catalogs along with the online ones.

Periodical databases were also created. For the most part, periodical databases contained information on the holdings rather than periodical literature indexes. Indeed, there are still no indexes to Vietnamese periodical literature. NACESTID is the largest producer of electronic databases, most of which center on scientific resources, in Vietnam. Its electronic databases are available on CD-ROMs as well as online.

The development of online catalogs and electronic resources offers Vietnamese library users quick access to information and unique collections. It also reduces the workload for catalogers. Vietnamese catalogers throughout the country now can rely on the database available at the NLV to catalog their materials. In fact, each month records generated by the NLV are uploaded on the network for catalog copying.

Though many Vietnamese libraries have automated many library functions, in general (with a few exceptions, e.g., the General Science Library [GSL], the second largest public library in Vietnam, which has used computers for cataloging new books since 1990), many of them continue to catalog materials manually. Furthermore, because of the lack of information networks, many online catalogs and electronic databases continue to run independently.

CD-ROM DATABASES

CD-ROMS made their way into the Vietnamese library communities in the early 1990s. Like PCs, the number of CD-ROM workstations, mostly stand-alone, in Vietnamese libraries was quite small. Major CD-ROM databases subscribed to by Vietnamese libraries include PASCAL, Chemical Abstracts, Derwent Biotechnology, Compendex, and IEEE/IEE Electronic Libraries.

The introduction of CD-ROM workstations and CD-ROM databases, however, has provided Vietnamese librarians with opportunities to acquaint themselves with the new form of library services that was about to dominate or transform not just the Vietnamese library services but also library services worldwide. Specifically, CD-ROM databases, unlike online electronic databases, which are expensive, provide low-cost electronic information sources. Another benefit in using CD-ROM databases, especially for the less technologically developed countries such as Vietnam, is that CD-ROM databases do not rely or depend on telecommunication facilities. Moreover, CD-ROM databases also provide low-risk experimentation with electronic resources. Indeed, the use of CD-ROM databases introduced many Vietnamese library professionals to electronic databases, particularly sophisticated searching techniques and data-extraction methods.

DEVELOPMENT OF LIBRARIES' LANS AND WANS

One of the first local-area networks (LANs) was established at the NACESTID and the NLV in 1994. Shortly after the development of the local area network, NACESTID created a wide-area network (WAN) in 1997. The newly established WAN enabled NACESTID to connect to information centers in sixty-one provinces. Most significantly, in 1997, NACESTID launched VISTA (Vietnam Information for Science and Technology Advancement)—a nationwide information network designed for resource sharing as well as meeting the information needs of the Vietnamese scientific communities and policy makers. Today VISTA serves as an Internet Content Provider as well as an Internet Service Provider. In fact, most universities and research and development institutions are VISTA contracted users. Compared to other Vietnamese libraries, NACESTID is the most technologically advanced information service entity in Vietnam. It owns many more electronic databases, either by purchasing databases from abroad or developing them itself, in the fields of science than any other libraries in Vietnam. In addition to NACESTID, which is under the management of MOSTE, other ministries, e.g., Ministry of Agriculture, have also developed their own LANs and WANs.

The NLV created its first local area network also in 1994. Shortly afterward, a wide-area network was established. The newly established WAN enabled the NLV to connect the NLV, the provincial libraries, and large municipal libraries together. Despite the desire of Vietnamese public library leaders, most Vietnamese public libraries, especially district and rural libraries, have yet to be connected to any of the national information networks.

In the area of higher education, by the late 1990s computers were also available in the libraries of Vietnam National University in Hanoi. However, it was not until the late 1990s that a LAN was established there. According to Mr. Nguyen Huy Chuong, director of the Library and Information Center of Vietnam National University at Hanoi, "All the departments for technical processing and circulation were equipped with computers which were connected to the main server in a hierarchical network. A multi-media service room was set up with eight networked multimedia PCs and other modern audio-video equipment" (Nguyen Huy Chuong 1998).

Also during this period, the University of Cantho, with foreign assistance (most of which came from the Netherlands), set up a local computer network in which the library's online catalog was installed. Over the years, foreign assistance has also enabled the library to acquire a good number of CD-ROM databases such as ERIC, Medline, and Derwent. Despite these developments, by the late 1990s and early 2000s, only large public and research libraries were connected to their "own" WANs.

THE INTERNET

The .vn domain was registered in 1994. However, it was not until December 1997 that Vietnam actually connected to the Internet. Since then, many of the governmental organizations, research institutions, and universities have connected to the Internet. For example, there were forty provincial and forty ministerial websites in 2003. Similarly, most university and research institutes have also established websites. To speed up Internet development in Vietnam, the government has given support to all research and development institutes, colleges and universities, and schools to develop "Internet stations."

Internet usage, however, is still low, even though Vietnamese do have access to the Internet, either through commercial, educational, government organizations, or Internet cafés. For example, in 2002 the percentage of Internet usage in Vietnam was 0.22 percent, compared with the world average of 9.10 percent.

Several factors contribute to this phenomenon. First, the government restricts the use of the Internet. Needless to say, Vietnam is still a communist state, in which the government places restrictions on what types of resources, regardless of formats, can be published, made available, or otherwise obtained by the public. Second, most resources in Vietnamese libraries are outdated. Moreover, a close examination of the resources available in Vietnamese libraries revealed that a substantial number of them are in foreign languages, such as English, French, or Russian, and most of them can be found in Western libraries, especially the United States or Europe. Third, most Vietnamese do not own either computers or telephones. For example, Vietnamese telecommunications officials hoped that by the end of 2004, telephone density would have reached 3.8 to 4 sets per 100 inhabitants and that almost all villages nationwide would have telephone services. Added to these issues, Internet fees (e.g., subscription, connection) are high. For the vast majority of Vietnamese, whose per capita income is about three hundred U.S. dollars, Internet usage is a luxury they simply cannot afford.

THE INTERNET AND LIBRARY SERVICES

Despite the limitations cited, Vietnamese library professionals have utilized the Internet to improve library services. In fact, many librarians searched the Internet for relevant and free resources and "added" them to their collections. The additions included free electronic journals, preprint journals, government documents, reference sources (dictionaries, encyclopedias), and e-books. For example, the library at the National University at Ho Chi Minh City has gathered and put to together an extensive list of "free" Internet resources, including free databases such as Asian Studies WWW Virtual Library, Medline, and dictionaries and encyclopedias such as *Botany Encyclopedia of Plants and Botanical Dictionary*, *Columbia Encyclopedia*, *Encyclopedia of Law and Economics*, and *Tech Encyclopedia*. Similarly, the library at the University of Cantho also maintains an extensive list of free Internet resources on its website. Included in this library's website are the websites that provide free but important electronic journals, such as *BioMed Central*, *PubMed*, and *PhysNet*. Many other libraries also located and provided URLs to governmental, international, and professional organizations or societies worldwide in which free professional or subject-oriented resources can be found.

Another important development with the introduction of the Internet into the Vietnamese library world is that many Vietnamese institutions, including libraries, have used the Internet to disseminate their electronic publications, especially journals, magazines, newsletters, and bulletins. Almost all major libraries in Vietnam publish at least a couple of electronic periodicals. NACESTID alone puts out over a dozen science-related electronic periodicals.

INFORMATION TECHNOLOGIES AND LIMITATIONS

Naturally, Vietnamese library leaders realize the importance of applying IT in all facets of library services and ultimately establishing a national information infrastructure that is capable of meeting the information needs of the Vietnamese as well as accessing global networked information resources. However, they also understand that these are monumental tasks, especially given the present economic and technological conditions in Vietnam.

It is an undeniable reality that it is costly to integrate IT into library services, from establishing and maintaining IT to using the Internet. The process entails a great many many costly elements, from buying and maintaining hardware, software, and networks to subscribing to and renewing electronic databases and other online services, to establishing and upgrading communications facilities, and hiring and training skilled information technology specialists (technicians and system engineers).

The most important element, of course, would be the financial resources to fund the other elements. Following the market-oriented transformation in 1986, the Vietnamese economy has improved significantly. Consequently, Vietnamese living standards have also arisen. However, Vietnam is still one of the poorest countries in the world, with annual income per capita about 300 dollars. To date, Vietnamese libraries have relied heavily on foreign assistance for the development of their IT. Indeed, although the Vietnamese government contributes to the effort, many foreign governments and organizations have furnished Vietnamese libraries with essentials ranging from hardware (e.g., computers, printers) to software (UN designed and donated CDC/ISIS software), to training for Vietnamese librarians in the application of IT into library services. Currently, the European Community is helping Vietnam to build a national portal for education, science, and technology. Similarly, the Swedish government is working to construct a scientific network called STENET, which aims at providing an environment in which information services and e-learning activities are offered.

Vietnamese libraries lack not only financial resources but also technical resources, especially the ability to develop information technologies—from software development to producing or training technical specialists (computers technician to systems engineers). There is a severe shortage of people with technology skills. Moreover, what makes this situation problematic is that people with technology skills prefer to work in the commercial sector, because the pay is much higher.

Another challenging aspect of IT is that there are only a handful of Vietnamese software developers or companies in Vietnam. It means that the Vietnamese continue to buy foreign IT, which are much more expensive. Besides the high cost of foreign IT, the lack of domestically designed and manufactured IT products makes it difficult for libraries to speed up the move toward application of IT to library services. Needless to say, it is more advantageous to have domestically designed and manufactured computer-related utilities since they can be modified to accommodate the libraries, especially with regard to Vietnamese language.

Added to these challenges is the fact that Vietnamese telecommunication infrastructures are still in the early stage of development. As noted above, most Vietnamese still do not own telephones or computers. As recently as 2004, there were fewer than four telephones per one hundred households. Moreover, as one expert noted, "The high cost of telephone charges and data services have turned many organizations away from any wide-area network (WAN) scheme, presenting major difficulties in the development of full Internet services infrastructure in Vietnam." Interestingly, the development of telecommunications infrastructures and facilities might slow down because of the penetration of cell phones in Vietnam. With cell phones, there is no urgent need to develop telecommunications infrastructures.

Finally, the lack of library standards, especially cataloguing standards, poses another monumental task for Vietnamese libraries. What has made the application of internationally recognized library standards problematic for Vietnamese libraries is that, because of the political situation in Vietnam over the last fifty years or so, Vietnamese libraries have used a variety of cataloging systems, from the American Dewey Decimal Classification (DDC) to the Russian Bibliographic Classification Scheme Bibliotechno Bibliograficheskaya Klassificatsiya (BBK), and might return to DDC. Vietnamese libraries are still examining the various cataloguing systems to see which one is suitable for Vietnamese libraries. Furthermore, currently there are no uniform fonts for Vietnamese characters.

"One expert points out that a significant outcome of the lack of a unified coding standard for the Vietnamese character set is that it is difficult to present Vietnamese information on the World Wide Web" (International Develeopment Research Center, 2002).

CONCLUSION

Vietnamese libraries have made significant progress toward the application of IT into library services. However, compared with libraries in developed countries, including the libraries of Vietnam's neighbors in the Association of Southeast Asian Nations, Vietnamese libraries technically still have a long way to go. Naturally, Vietnam needs to build a national information infrastructure, capable of meeting the information needs of all Vietnamese, from peasants to policy makers to scholars, as well as sharing its resources worldwide. In conjunction with the technical development aspect of library services, Vietnam, however, should also pay attention the development of the other vital elements of library services, which at the present time are unmistakably inadequate. They include the need to build up library sources for all libraries, train and increase the number of library professionals and staff, devise library standards, erect modern library facilities, play a greater role in the curriculum development at all levels of education, develop information sharing services, and articulate information needs among users.

REFERENCES

Cao, Minh Kiem. 2000 (October). "Integrating Electronic Resources in Traditional Library Services at the National Center for Scientific and technological Information and Documentation." Paper presented at the Special Library Association Global Conference, Brighton, United Kingdom.

Clow, David. 1996. "Vietnam Library Update with Special Reference to Scientific and Technical Library Network." *Libri* 46, no l: 25–34.

D'Amicantonio, John. 1998. "Special Library Experience in Vietnam." *Information Outlook* 2, no. 12: 8–10.

Dao, Tien. 1996 (July). "Networking in Vietnam." *Provenance* 1, no. 3: 15–18.

Diep, Kim Chi, and Huynh Thi Trang. 1998. "A History of the Adoption of New Information Technologies by Cantho University Libraries." Pp. 35–42 in *NIT' 98: The 10th International Conference for New Information Technology*. West Newton, MA: MicroUse Information.

Gooch, Peter. 1994 (October–December). "Information Flows in Agricultural Research in Vietnam: Status and Prospects." *Quarterly Bulletin of the International Association of Agricultural Information Specialists* 39, no. 4: 312–18.

International Development Research Center. 2002. Grassroots Networking in Vietnam," http://web.idrc.ca/en/en/ev-10126–1-do_topic.html (accessed 2 November 2004).

Jarvis, Helen. 1987 (September). "Microcomputers in Vietnam 1987: With Particular Reference to the Vietnam Union Catalog (VUC)." *Microcomputers for Information Management* 4, no. 3: 173–81.

Kieu, Van Hot. 1998. "Information Technology in Public Library System of Vietnam." Pp. 83–86 in *NIT' 98: The 10th International Conference on New Information Technology*, edited by Cheng-chin Chen. West Newton, MA: MicroUse Information.

MacMillan, Sandy. 1990 (December). "Library and Information Services in Vietnam Today: The Basis of Development." *Libri* 40, no. 4: 295–305.

National Library of Vietnam. www.nlv.gov.vn/english/inc/head.htm

Nguyen, Huy Chuong. 1998 (November). "Automating Vietnam's Academic Libraries: The Example of Vietnam National University." *Asian Libraries* 7, no. 11: 333–38.

Nguyen, Huy Chuong, and Ton Quoc Binh. 1998. "A Model for Automating the Library and Information Services at the Vietnam National University, Hanoi." 185–92 in *NIT' 98: The 10th International Conference on New Information Technology*, edited by Cheng-chin Chen. West Newton, MA: MicroUse Information.

Nguyen, Minh Hiep, and Hoang Le Minh. 1998. "Building the Library System of National University of Ho Chi Minh." 193–98 in *NIT' 98: The 10th International Conference on New Information Technology*, edited by Cheng-chin Chen. West Newton, MA: MicroUse Information.

Nguyen, Thi Bac. 1998. "The General Sciences Library of Ho Chi Minh City: Technology in Work Improvement." Pp. 199–204 in *NIT' 98: The 10th International Conference on New Information Technology*, edited by Cheng-chin Chen. West Newton, MA: MicroUse Information.

Nguyen, Thien Can. 1998. "Issues in Automation Facing Private University and Special Libraries: A Case Study of Two Libraries in Ho Chi Minh City." Pp. 205–10 in *NIT' 98: The 10th International Conference on New Information Technology*, edited by Cheng-chin Chen. West Newton, MA: MicroUse Information.

Pham, The Khang. 2001 (August). "Mobile Libraries in Vietnam in 21st Century." Paper presented at the 67th International Federation of Library Association Conference, Boston, Massachusetts.

Quarterman, John S. 1998 (April). "The Internet in Vietnam." *SunExpert Magazine* 9, no. 4: 56–60.

Ta, Ba Hung. "Bridging the Digital Divide in Vietnam." www.stii.dost.gov.ph/astinfo2/frame/july_to_sep_2k1/pg_10_to14.htm

"Telecoms Sector to Double by 2010." www.vista.gov.vn/vistaenglish

Tran, Lan Anh. 1999 (January). "Recent Library Developments in Vietnam." *Asian Libraries* 8, no. 1: 5–16.

———. 1999 (October). "The Implementation of Information Technology in Vietnamese Libraries: Results of a Survey." *Asian Libraries* 8, no. 10: 380–95.

Tran, Manh Tuan. 2003 (October). "S & T Information Activities in Vietnam: Situations and Problems." Paper presented at the 5th Co-Exist-SEA Workshop, Manila, Philippines.

Tran, Ngan Hoa. 1999 (June). "Providing Agricultural Information to Decision Makers and Researchers in Vietnam." *Quarterly Bulletin of the International Association of Agricultural Information Specialists* 44, nos. 2 & 3: 134–35.

Trung Tam Thong Tin Tu Lieu Khoa Hoc va Cong Nghe Quoc Gia (National Center for Scientific and Technological Information and Documentation). www.vista.gov.vn/english/main/main.htm.

United Nations Development Programs. "VARENET as the Pioneering Effort." http://undp.org.vn/themes/ict4d/annerstedt/it2k09b.htm

Vu, Van Son. 1993. "Libraries and Documentation in Vietnam, with Special Reference to Science and Technology." Pp. 80–89. in *Information and Libraries in the Developing World*, edited by A. Olden and M. Wise. London: Library Association.

———. "Electronic Libraries: Feasibilities of Their Building Up in Vietnam." www.lub.lu.se/sida/papers/son.html

World Economic Forum. 2004. The Global Information Technology Report, 2003–2004: Towards an Equitable Information Society. New York: Oxford University Press.

Chapter 5

The Impact of Technology on Chinese Library Collections and Services

Jing Liu and Ian Yiliang Song

ABSTRACT

Information technology (IT) has brought dramatic changes to Chinese libraries in the past decades, fundamentally altering libraries' main function from preservation to access. This chapter gives a brief overview of the history of Chinese libraries, introduces technological developments and applications, and discusses their impact on collections and information services. Major difficulties and challenges in Chinese libraries have been raised on the basis of observations and a literature review.

INTRODUCTION

With a splendid history and treasured collections, Chinese libraries have been going through dramatic changes. Especially in the last decade, Chinese librarians have made great efforts to catch up with the unprecedented opportunity technology has brought to improve their collections and services. Having worked in libraries in China, the United States, and Canada, the authors are especially impressed by the progress Chinese libraries have made in applying new information technologies to library operations. The long-standing gap between Chinese and the Western libraries has been greatly reduced. This chapter will introduce an overview of library development in China and discuss recent technological applications and their impact on collections and information services. The discussion is based on firsthand experience and observation, as well as literature in both English and Chinese.

LIBRARIES IN ANCIENT CHINA

Inscriptions on bones or tortoise shells are the earliest writing records found in China, dated to the fourteenth century BC. Collecting written records was originally undertaken by the government and aristocracy. Confucius (551–479 BC) compiled the records and archives available and made them into "books" on bound wood or bamboo sticks. Among the Hundred Schools of Thought, the Confucian school became the basis for the order of Chinese society. In the West Han Dynasty (206 BC–AD 24), books were well organized and recorded as "thousands of volumes in 72 categories and 900 schools."[1] Invention of papermaking and printing made it possible to produce books large scale, with multiple copies. Printed books emerged around the ninth century in China. With the invention of movable type in the Song Dynasty (AD 960–1279), book printing prospered. It produced not only a large numbers of books, but books of high quality.

Through dynastic Chinese history, books were stored in imperial storage buildings, Buddhist temples, schools, and private houses. They were cataloged and classified into four categories: classics, history, philosophy, and arts. Those book storage buildings can be considered early libraries in China. Some private and school collections not only lent their books but also made duplicates and compiled bibliographies. "In the 16th and 17th centuries most people could afford to buy books that were produced for readers with different levels of income."[2] Private collections grew rapidly in number and size. Tianyi Library, built in the mid-sixteenth century, still exists in Ningbo, Zhejiang Province, with a collection of 300,000 volumes. The emperors of the Ming and Qing dynasties distributed books as gifts to local governments. They gathered rare books and gave orders to carry out huge compilation projects in which books were carefully selected from the imperial collections. The *Yongle Da Dian* (Yongle Canon), compiled in the Ming Dynasty between AD 1403 and 1407, contained 22,877 volumes in 11,095 books. "It was 12 times that of the famous encyclopedia compiled by the French author Diderot in the 18th century."[3] *Encyclopedia Sinica (Si Ku Quan Shu)* was a large series of books, compiled by scholars under the auspices of the Qing government during the reign of Emperor Qianlong (1735–1796).

At the beginning of the twentieth century, imperial rule in China came to an end. "Bourgeois political reform movement introduces Western library theory and practice, and encourages the establishment of provincial libraries."[4] Mary Elizabeth Wood (1862–1931), a trained librarian at Columbia University, was one of the most outstanding contributors to Chinese librarianship. She established the first library school and sent the first group of Chinese students to the United States for professional training; they later assisted her in promoting library services and librarianship in China. According to *Shen Bao Year Book,* there were 5,828 libraries across the country in 1936,[5] but after years of suffering from wars, fewer than four hundred libraries survived when the People's Republic of China (PRC) was established in 1949.[6]

LIBRARIES IN THE PEOPLE'S REPUBLIC OF CHINA

A new library system was gradually formed within the PRC administrative structure. Large collections (over a million volumes) were built in the following systems:

- Public libraries, such as the *National Library, Shanghai Library* and provincial libraries throughout the country under the Ministry of Culture. "They are self-governing at the various levels of local authorities: province, county, city, and township."[7]
- Research libraries, such as the libraries of the Chinese Academy of Science (CAS) and the Chinese Academy of Social Sciences (CASS), and their thousands of regional branches throughout the country
- Academic libraries owned by universities and colleges, under the Ministry of Education

By 1996, China had 250,000 libraries with a cumulative collection of three billion volumes. Among them, 2,615 were provincial and city public libraries, 1,080 university libraries, 8,000 research libraries at all levels, and more school and special libraries.[8] Despite the damage and hindrance caused by the Cultural Revolution (1966 to 1976), Mr. Huang Zongzhong, a library expert, believes library "expansion in numbers and space is one of the biggest achievements in the new China."[9] Meanwhile, two major events changed the path of library development and sped up the transition from the traditional to modern libraries. One was the development of computer coding system for Chinese, so that computers can process information in Chinese characters; another was when Chinese MARC (CNMARC) was released in 1990 and distributed by the National Library for use in all library systems.

In the past decade, China has entered a period of rapid economic growth and joined the global IT industry. The central government has invested intensively in national information infrastructure. CHINAPAC and CERNET (China Education and Research Network) were the earliest networks that emerged in the mid-1990s. Improvements in living conditions have resulted in hunger for knowledge and education; IT has allowed much greater access to an expanding world of information; "Over 87 million users in the mainland of China utilize the information on the Internet in June 2004"[10]and a year later in 2005, the Internet users number has reached 103 million."[11]

THE DEVELOPMENT OF AUTOMATED LIBRARY SYSTEMS

Librarians in China have been paying close attention to the development of new technology, especially computer and communication technologies. It was a dream of all the professional librarians, inspired by the Chinese government's "Four Modernizations" plan, to accomplish library automation in the 1980s and early 1990s. They were eager to learn about automated library systems from advanced countries and "anxious to consider possibilities for networking and cooperation on regional, national and international levels."[12] However, it has been a long journey to make this dream come true.

In the mid-1990s, many universities, libraries, and other academic institutions were trying to develop their own systems based on local needs, without common standards. The result was chaos in the late 1990s and early 2000s, when a large number of universities and colleges merged.[13] Libraries had too many choices in choosing their systems, and the market for the library systems was very competitive. Some successful domestic integrated systems have been widely adopted by academic and public libraries. ILAS (Integrated Library Automation System) version 5.0, developed by Shenzhen Library, has six different versions for academic, public, corporate, compact, and traditional libraries, and has been used in 1,300 libraries. The Libsys2000, developed by Jiangsu Huiwen Software, is another successful system that has been widely used. Libraries with strong financial support also tried alternative foreign-made systems. For example, the National Library selected Aleph 500 by ExLibris, and Peking University adopted SIRSI, which was enhanced on campus with more Chinese features.

After a decade of development there are 15,437 libraries in China with automated library systems, according to the *Research Report on the Construction of Library Systems* in 2004, including 2,697 public libraries, 1,700 university/college libraries, 4,100 research libraries, 1,600 party member training school libraries, 1,200 hospital libraries, 3,800 technical secondary school libraries, and 340 other libraries. Seventy-six percent of libraries use their systems only on parts of library management (such as cataloging); 18

percent of libraries use integrated systems that are applied library management as a whole; 6 percent of libraries implemented integrated systems and also provide more comprehensive services.[14]

THE DEVELOPMENT OF
INTERNET-RELATED TECHNOLOGIES

China sent out its first e-mail in 1987, and the NCFC (National Computation Facilities of China) Project, funded by the World Bank, was the first to make the Internet become a reality in China in1994.[15] The CERNET, funded by the Chinese government, was the first national education and research computer network; it had eight regional centers and covered more than 100 university campus networks in 1995. As of 2003, ten regional centers with more than 1,300 universities and research institutions were connected to the CERNET, as well as more than 15,000,000 users. The next generation of the Internet, IPv6 (Internet Protocol version 6) was implemented in the CERNET in 2004, when IPv6 replaced IPv4. CERNET 2 now can provide high-speed Internet service at 2.5 gbps–10 gbps.

Librarians in China showed great enthusiasm for the Internet and tried to find potential applications of Internet-related technologies in libraries. The introduction of the Internet also created enormous concern over libraries' and librarians' roles in the emerging technologies and new information infrastructure.[16] It is now common for libraries to create their own portals or websites in order to provide users with more services, such as user guides, Web-based OPACs, database access, virtual reference, and useful Web links. Limited sources and information makes some library sites less attractive, however, and some library systems cannot be integrated into the World Wide Web environment, leaving users without access to library catalogs.[17]

LIBRARY COLLECTIONS

Print Collections

Libraries have suffered two down periods in the history of PRC. The first, during the Cultural Revolution catastrophe, left the professionals with a ten-year collection gap. The second was during the market economic transition period. Then in the 1980s the publishing industry started booming. A 1997 article reported that "China ranks No.1 in the world in total number of annual publications,"[18] with "101,381 monographic titles, 7,583 periodicals and 2,089 newspapers" published in 1995.[19] In contrast, all types of libraries were badly in need of new books. Compared to soaring book prices, library funding was much lower. Public library acquisition rates decreased by 9.2 percent every year from 1990 to 1995. Among the 2,500 public libraries, 566 purchased no new books in 1995.[20] Academic libraries' situation was not much better. The number of newly acquired items went down sharply, and they had to stop subscribing to some Western academic journals. The difficulty of developing collections forced librarians to seek other ways to meet patrons' needs, such as technology, cooperation, and resource sharing.

At the same time, computers, multimedia, and CD-ROMs were gaining popularity, especially in academic and large municipal libraries. Academic libraries started providing access to DIALOG and began to use full-text CD-ROMs purchased abroad. Access was limited to individual libraries and often restricted to elite users only. Librarians desperately needed to utilize emerging technologies to enhance their collections.

Electronic Resources

In the late 1990s, administrative authorities at all levels began to conceptualize the Internet as a vehicle for growth and put the building of digital libraries on their agenda. Libraries were getting more attention and support. Those with unique and rare collections started compiling bibliographies and indexes in print and electronic formats. Some even started digitizing entire collections with technical partners. Many Chinese librarians heard the word "digital library" for the first time in a special session: "Digital Libraries, Technologies and Organizational Impacts" during the 62nd IFLA General Conference in 1996 held in Beijing.[21] People did not seem to care much about reaching a unified definition of "digital library" but acted fast to digitize large amount of materials in Chinese. In April 2000, the China Digital Library Project was launched, and it "started scanning 200,000 pages per day."[22]

Large numbers of Chinese e-books are integrated into individual library collections across the country. A single interfaced search engine—Duxiu—just came live at www.duxiu.com and covers over 1,800,000 Chinese titles, of which many have their digital versions available from a variety of holders. Duxiu allows users to search bibliographic information, tables of contents of a book, and even some full-text. It has other functions such as trial reads, MARC records, and links to suppliers. It can hold its own when compared with Amazon or Google, especially in terms of providing bibliographic information.

Due to reasonable price and convenience of access, e-books quickly entered library collections. Libraries provide their card holders access via mirror websites. Library collecting e-books in China can be categorized into the following types:

- Chinese Classics and Dictionaries (by Unihan) (www.unihan.com.cn). For example, *Encyclopedia Sinica* was built beautifully into a full-text database, which contains full-text and scanned images of some 3,700 titles. The new features, including full-text searching, annotation, and hyperlinks in the text, make it a great research tool and gained popularity among Chinese libraries. This kind of digital collection is even welcomed by Japanese researchers and warmly embraced on the Web by scholars in the West as well.
- Beijing SuperStar's (www.ssreader.com) e-books cover all subjects. Many archives, research institutions, and public or academic libraries worked with Superstar to digitize their entire collections. Ms. Liu visited the Beijing-based company in October 2004 with a group of Chinese Studies librarians, who were impressed by the demonstrations and introductions of the e-book collection. The U.S. Library of Congress and the University of California, San Diego, are the first two institutional customers in North America.
- Also notable are Apabi (www.apabi.com) and The China Scholars' Home (www.21dmedica.com), which offer born-digital and digitized e-books to libraries. Libraries can subscribe to Apabi's e-book collection just as they can to *NetLibrary*'s English e-books, the service it is modeled on.

Besides e-books, e-journals are also widely available from three major providers. Among them, VIP (www.tydata.com) from Chongqing, Sichuan Province, provides over 12,000 e-journals back to 1989. TTOD (www.cnki.net) provides 5,300 academic journals from 1994 on and updates full-text articles daily. The latest e-journal provider, Wanfang (www.periodicals.com.cn), provides a large number of digital dissertations and has flexible subscription and payment options that quickly attracted the library market. The above-mentioned digital content providers occupied the entire library market. They united libraries to enrich digital e-sources with strong technical support.

On top of millions of e-books and thousands of e-journals, domestic databases are developing so fast that research and academic libraries are getting more and more interested in them. Chinese Science Citation Index Database, Chinese Science Document Database, and Chinese Social Sciences Citation Index are the top general databases. China Infobank offers a Web-based online service with access to a number of full-text Chinese language databases on news, business, legal, and statistical information. Some databases for unique Chinese heritage resources, such as the Chinese Ancient

Text (CHANT) developed by scholars in Hong Kong and Mainland China, showed the world of the oracle and bronze inscriptions in digital format for the first time. Many technically strong libraries went even further to create, collect, and provide access to local or institutional digitized information. For instance, Shanghai Library (www.library.sh.cn) started with a digital conversion of its own rare books in 1997. It finished "seven digitization projects and accumulated 200 GB of its own digital resources in multimedia format by 1999."[23] As individual libraries' digitization projects developed their own portals, thousands of specialized and localized new databases joined the National Cultural Resource Sharing Project.

Coordinated Purchasing

China has a unique national library system. Each type of library has separate administration and collection strengths though they traditionally have lacked cooperation. New computer network technology links many libraries together and offers a platform for coordinated purchasing and resource sharing. For example, academic libraries in the Shanghai area have started coordinated acquisition of Chinese full-text databases. Now CALIS operates four national subject information centers that are in charge of coordinated purchasing. By the year 2001, 251 university libraries had joined the network and were able to share their bibliographic and holding information. Since 2002, dozens of major universities have been working together to acquire foreign serials and electronic resources. For instance, CALIS members purchased Academic Search Premier and Business Source Premier in a coordinated manner. Each member in CALIS benefits from the coordinated purchase, and even the large libraries that paid the largest share in the coordinated purchase still saved 30 to 40 percent of the cost of subscribing alone. Between 1997 and 2003, 500 libraries worked together in negotiating and licensing more than 120 databases and 12,000 e-journals in foreign languages.[24]

National Cultural Resource Sharing Project (www.ndcnc.gov.cn)

The National Cultural Resource Sharing Project (NCRSP), often called "the Sharing Project," was announced in April 2002 by Xinhua News Agency. The network, to be completed in 2005, intends to enable people in remote and poor western areas to share the cultural resources in libraries, museums, art galleries, and research institutes in the affluent east. "At the end of April 2003, the Sharing Project could provide access to 1.4 TB of digital resources, including 100 lectures given by prominent experts and scholars; 660 movies; 132 regional plays; 320,000 images; 80,000 books;

460,000 periodical titles; and rich knowledge of history, law, popular science, healthcare, and life."[25] Libraries in this project play a crucial role.

INFORMATION SERVICES

In China, the development of computer and communication technology (ICT) has challenged traditional library services just as it has elsewhere in the world. The most important change ICT brings to libraries is from the traditional library service that users have been familiar with to a totally new type of service, due to the change in the nature of library's collection and information delivery. The library services become more computer dependent, more user oriented, and more interactive, as we detail in the following sections.

Online Aggregate Database Services

In 1983, ISTIC (Institute of Scientific and Technological Information of China) gained access to DIALOG and STN, a great development in the history of information services in China. Most Chinese librarians and end users had never seen a computer or had any experience using online databases. What the end users could do was fill out a search request form and provide some keywords, and trained "specialists" searched databases for them. The cost was one of the major concerns for users, as researchers could not afford the searches without research funding.

In the late 1980s and early 1990s, more affordable foreign databases on CD-ROM were available to Chinese users and libraries. For example, most medical universities/colleges and some hospital libraries purchased access to Medline on CD-ROM from different vendors (such as Silver Platter and Cambridge). One 286 IBM-compatible computer with a 30 MB hard drive, 64 KB RAM, a 5½-inch floppy drive, and one Hitachi single-speed CD-ROM drive under MS-DOS 3.0 was everything needed for a CD-ROM workstation.

From the late 1990s, World Wide Web technologies began to dominate the field of information services. Chinese information makers and database vendors kept pace with the world. Many domestic products, such as those from Wanfang Data, VIP Information, and CNKI (China National Knowledge Infrastructure), become an important part of the information market.[26]

Library Instruction

More and more databases are available to Chinese libraries and their users. Users feel both excited about many databases that they never used before and frustrated because different interfaces and searching facilities from different databases make using them difficult. To meet users' needs, Chinese libraries of all types offer free courses to their users. Librarians are playing an important role in information literacy. Chinese academic libraries began offering information literacy courses to their students and faculty members in the 1980s, but now they focus more on searching electronic databases, computer literacy, and the Internet. "Literature Searching and Utilization" courses are required for undergraduate and graduate students in Chinese universities.[27] Chinese public libraries also offer different training courses to meet the needs of different user groups. For example, Xiamen Public library offered training sessions for users on Internet searching, e-mail, Telnet, and database searching.[28]

Virtual Reference and Document Delivery

Reference service has been a weak point in the Chinese library community throughout its history. Liao Jing viewed the problem as rooted in "the millennia-old public-phobic practices of imperial and private libraries."[29] Librarians used to work "behind the scenes" and focused on the particular a group of clientele. Network applications seem to be able to improve reference service technically. While in-person reference services are still limited and poor, virtual reference and document delivery, on the other hand, have made considerable progress.

Sun Yat-Sen Guangdong Provincial Library (www.zslib.com.cn), located in the economically advanced Zhujiang River Delta, started using an in-house-developed integrated system in 1991. It has access to over 400,000 e-books and a hundred databases from a variety of digital content providers. Every month over 100 million pages within its digital collection are accessed on the Web, a rate ten times higher than for its print collection.[30] Its BBS-based virtual reference service has twelve reference librarians and ten volunteers, who can be anywhere in the world. The service is free and available twenty-four hours a day, from 8:00 AM to 9:00 PM monitored by regular librarians, and monitored the rest of time by volunteers. It answered 150 questions per day on average and delivered 1,500 electronic items to users' desktops in 2002. Library users are all over the world. During the SARS (severe acute respiratory syndrome) outbreak in 2003, many libraries closed or reduced opening hours, but people could still benefit from this service. Today, Sun Yat-Sen Guangdong Provincial Library is leading a collaborative virtual reference service that is part of the national sharing project, providing information by phone, e-mail, and online via IM and SMS.

In the academic and research library community, instead of e-mail or BBS, Question Point was adopted and provides real-time interactive reference. CALIS members are planning a collaborative virtual reference service to serve university users.[31]

Scientific and Technological Project Search Service

Scientific and Technological Project Search Service is an in-depth reference and information service. This is a service unique to Chinese libraries, and the goal is to provide comprehensive information searches to meet the background needs of a particular project.[32] This service started in medical universities/colleges in 1985.[33] It is also called "novelty search," a term that is probably not familiar to most people outside China but has been used in the library and research community in China for several years. The service is to check, by searching electronic databases and looking up print materials, the state of the art of a particular research field to see whether a particular project has been done, what has been done, how it has been done, and how many research articles or other materials are related to this subject. In China, scientists or researchers who apply for research funding, grants, patent registration, or academic awards are requested to submit their applications with Scientific and Technological Project Search Service reports. The Ministry of Education of China issued the regulations about the service in 2001,[34] and the authority to conduct novelty search service has been given to forty-three academic libraries. [35] Most academic libraries do not have that service, which is offered as a fee-based reference service.

Computer Lab of Electronic Materials

With the changes in libraries' collection, more and more electronic materials on CDs, CD-ROMs, DVDs, and multimedia materials on local computer networks and the Internet are available to library users. Most academic and public libraries have computer labs for these materials. It has become the most valued place in a library. Users can browse the Internet, search databases, learn computer software, listen to music, and watch movies. Unfortunately, users have to pay for the service.[36] It is an obvious trend to provide library users with these materials in libraries.

CHALLENGES

The Digital Divide

The collections and services introduced in this article are concentrated on the major national public, academic, and research libraries. The development of library collections and services in China is uneven. The difference between information-rich and information-poor becomes more obvious than ever. Urban elites embrace the new IT and enjoy improved information services, while the majority of people who inhabit the vast rural areas are left behind. It is "accelerating with full steam on top of the traditional informational divide as long as the expressed concerns stay as rhetoric only."[37]

Information Freedom and Legal Issues

The Chinese government has been trying to control information on the Internet from within Mainland China, as well as to restrict public access to sensitive political information from outside mainland China. The policy has a great impact on Internet access.[38]

In contrast, intellectual property law did not give clear explanations on digitization and dissemination on the Internet until October of 2001. The legal expert, Mr. Chen Chuanfu, suggested that librarians' legal knowledge is the key to protecting authors' rights.[39] In June 2002, Professor Chen Xingliang of Peking University won the first copyright lawsuit against the China Digital Library.[40] This case sent out a clear warning to all libraries in China that they need to comply with the law even if they are not trying to profit by their actions. The case caused heated debate on the inevitable but seemingly elusive digital future. Now the major digital content providers that were mentioned above are actively seeking permission from individual authors. They are also working with the National Copyright Protection Center on a collective license program. The fees charged to users are used to pay for this collective license. Balancing the interests of the owner, user, and society will be crucial to the growth of the digital libraries in China.

Overlapped Collections and Lack of Cooperation

A national library resources survey was conducted in the early 1990s.[41] The result showed that the major collections were highly localized in areas like Beijing and Shanghai and that the rate of overlapping was shocking. Digital library development has the same problem. This has been one of the major concerns for the profession, and cooperation on all levels is desperately needed. In the new millennium, information resource sharing mechanisms are being developed. Three national projects have been instituted to share information resources: the China Advanced Education Library Information System, sponsored by the Ministry of Education; the China Library Information Network, sponsored by the Ministry of Culture; and the National Science and Technology Library, sponsored by the Ministry of Science and Technology.[42]

CONCLUSION

Libraries in China hold treasures from imperial and private collections that have existed for thousands of years. In the past twenty years, technology has brought tremendous power to Chinese libraries and changed library collections and services dramatically. Empowered by the new IT, librarians are able to provide enriched information resources in all formats and to serve users more efficiently. Library work as a whole has been tremendously improved in the past decade. Librarians have shown that they are the crucial players in the national information revolution. Despite unbalanced progress, the most obvious achievements are networking and large-scale digital collections. Reference services and user education are gaining in importance. Regional and academic library cooperative systems are gradually forming. Like elsewhere in the world, the professionals in China are still in the process of discovering what to do with new technology in information services. As the nation grows toward a knowledge-based economy, its libraries will certainly continue to make progress.

NOTES

1. Lai Xinxia, *Zhongguo gu dai tu shu shi ye shi* [Book History of Ancient China], (Shanghai: People's Publishing House, 1990), 28.

2. Chow Kai-wing, *Publishing, Culture, and Power in Early Modern China* (Stanford: Stanford University Press, 2004), 55.

3. Xinhua News Agency, "China Publishes Photocopies of World's Earliest Encyclopedia," available from www.china.org.cn/english/2002/Apr/30967.htm; (accessed January 16, 2005).

4. Gong Yitai and G. E. Gorman, "Reader and Information Services," in *Libraries and Information Services in China* (Lanham, MD: Scarecrow, 2000), 129–30.

5. *Shen bao nian jian [Shen Bao Yearbook]* (Shanghai: Shen Bao Guan, 1937), 1235.

6. Huang Zongzhong, "New China Library Work 50 Years," in *China Library Yearbook 2001* (Beijing: National Library Press, 2001), 29–32.

7. D. J. Foskett, "Chinese Libraries in the 1990s: A Western View," *Asian Libraries* 8, Cumulative Issue (1999): 241–49.

8. *China Library Yearbook 1996* (Beijing: National Library Press, 1997), 11.

9. *China Library Yearbook 1996* (Beijing: National Library Press, 1997), 6.

10. China Internet Information Center, "The 14th Statistical Survey Report on the Internet Development in China, July 2004," available from www.cnnic.net.cn/download/2004/2004072003.pdf (accessed January 2005).

11. People's Daily Online, "A Digital China in the Eyes of Foreigners," available from http://english.peopledaily.com.cn/200412/22/print20041222_168283.html (accessed January 28, 2005).

12. Robert D. Stueart, Margo Crist, and Barbara I. Ford, "China's Libraries in Transition," *Library Journal* (12 September 1987): 143–47.

13. Huang Hong-Wei and Zhang Sha Li, "College and University Mergers: Impact on Academic Libraries in China," *College & Research Libraries* 61, no. 2 (2000): 121–25.

14. China Information Industry, "Tu shu guan xin xi hua jian she zong he bao gao (2004)"[Research Report on the Construction of Libraries Systems (2004)], available from http://txzxs.cnii.com.cn/20030527/ca262582.htm (accessed January 17, 2005).

15. Zhu Qian, "Latest Development of Internet in Mainland China," available from http://lark.cc.ku.edu/~eastasia/paper01.html (accessed January 15, 2005).

16. Wen Shufang, "Qian tan xin xi gao su gong lu dui tu shu guan de ying xiang" [Information Highway Influence on Libraries], available from www.fslib .com.cn/article/view.asp?id=1996032311 (accessed January 15, 2005).

17. Liu Peijun, "Tu shu guan wang zhan de xian zhuang ji fa zhan dui ce" [The Current Situation of Library Web Sites and Perspective Solutions], available from www.fslib.com.cn/article/view.asp?id=2001121251 (accessed January 16, 2005).

18. Zhou Yuan, "Catching up with Technology: Recent Developments in Chinese Libraries," *CALA E-Journal*, no. 11 (November 1997), available from www .white-clouds.com/cala/publications/e-journal/calaej11.htm (accessed January 28, 2005).

19. *1995 nian quan guo tu shu, za zhi, bao zhi chu ban tong ji zi liao* [1995 National Statistics for Books, Periodicals and Newspapers Publishing] (Beijing: Zhongguo tong ji chu ban she, 1996), 1–4.

20. 1995 National Statistics for Books, Periodicals, and Newspapers Publishing (Beijing: Zhongguo tong ji chu ban she, 1996), 6.

21. Liu Wei, "New Development of Digital Library in China," available from www.kc.tsukuba.ac.jp/dlkc/ e-proceedings/papers/dlkc04pp120.pdf; Internet (accessed January 28, 2005).

22. Shen Difei, "Lun dang qian wo guo tu shu guan zi dong hua" [Current Library Automation], *China Library Yearbook 2001* (Beijing: the National Library Press, 2001), 13–17.

23. Liu Wei, "The Shanghai Digital Library Initiative," *Asian Libraries* 8, Cumulative Issue (1999): 262– 65.

24. Xiao Long andYao Xiaoxia, "Wo guo tu shu guan dian zi zi yuan ji tuan cai gou mo shi yan jiu," [Library Electronic Resources Coordinated Purchasing] *Zhongguo Tu Shu Guan Xue Bao*, no. 5 (May 2004): 31–34.

25. Wang Fenlin, "A Summary: Digital Resources Creation and Services of National Cultural Resources Sharing Project," available from www.ndl.go.jp/en/ publication/cdnlao/047/472.html (accessed January 17, 2005).

26. Gu Zhenyu, "Lian ji jian suo xi tong" [Online Information Retrieval Systems], available from www.istis.sh.cn/zyjc/zysj/dbprofile4.pdf (accessed January 17, 2005).

27. Wu Jiangzhong and Huang Ruhua, "The Academic Library Development in China," *Journal of Academic Librarianship* 29, no. 4 (2003): 249–53.

28. Zeng Dexing," Wang luo huan jing xia gong gong tu shu guan yong hu xu qiou fen xi" [Analysis on the End Users of Public Libraries in the Internet Era], available from http://210.34.4.20/society/jqli/society/paper/2003meeting/zxdwlh .htm (accessed January 18, 2005).

29. Liao Jing, "A Historical Perspective: The Root Cause for the Underdevelopment of User Services in Chinese Academic Libraries," *Journal of Academic Librarianship* 30, no. 2 (2004): 109–15.

30. Mo Shaoqiang and Tan Zhichao, "Digital Reference Service: A Practice and a Study," *Knowledge Navigation and Library Services: The Proceedings of the First Shanghai International Library Forum* (Shanghai: Scientific and Technological Literature Publishing House, 2002), 156–59.

31. Zhang Aiyou, , "Lun jian li gao xiao tu shu guan lian he can kao zi xun fu wu xi tong" [Establish Academic Library Collaborative Reference System], *Tu Shu Qing Bao Zhi Shi*, no. 1 (2004): 68–70.

32. Yitai Gong and G. E. Gorman, "Reader and Information Services," in *Libraries and Information Services in China* (Lanham, MD: Scarecrow, 2000), 129–30.

33. Liao Ximing, "A Summary of Retrieval of Search for the Latest S&T Achievements," *Information Research*, no. 1 (March 2003): 1–3.

34. China Education and Research Network, "Ke ji cha xin ji gou guan li ban fa," available from www.edu.cn/20010821/189542.shtml (accessed January 24, 2005).

35. Science and Technology Development Center, "Jiao yu bu ke ji cha xin gong zuo zhan," available from www.cutech.edu.cn/kejichaxin/default.html (accessed January 24, 2005).

36. Chen Xiupin and Zou Wu, "Service Functions of Electronic-Reading Rooms," *Information Research*, no. 1 (March 2003): 4–5

37. Andrew Lew, "Turning Digital Divide into Digital Combine: Effecting Public Digital Library Systems to Serve the Entire Population of China," in *City Development and Library Services: The Proceedings of the Second Shanghai International Library Forum* (Shanghai: Scientific and Technological Literature Publishing House, 2004), 275–81.

38. Wilfred W. Fong, "The Development of Library and Information Technologies in Southeast Asia," *Information Technology and Libraries* 16, no. 1 (1997): 20–26.

39. Chen Chuanfu, "Guan cang wen xian shu zi hua de zhi shi chan quan feng xian," [Analysis of Intellectual Property Risk in the Procession of Digitalizing Library Collection], *Tu Shu Qing Bao Zhi Shi*, no. 5 (October 2003): 2–5.

40. *Gazette of the Supreme People's Court*, no. 2 (2003): 34–35.

41. Wu Weici, "ZhongguoTu shu guan shi ye fa zhan li cheng," in *China Library Yearbook* (1996), 20–55.

42. Li Jing and Meng Liansheng, "Developing Scientific and Technological Information Sharing in China," *Information Development* 19, no. 3 (2003): 179–81.

Chapter 6

Information Technologies and the Development of South Asian Collections and Services in U.S. Libraries

Rajwant S. Chilana

ABSTRACT

This chapter discusses the development of South Asian collections and services in U.S. libraries over the years and how it has been influenced by emerging information technologies. Beginning from the development of microforms to the current trend of online resources, technology has been used in various ways to enhance information storage and retrieval. The chapter focuses on some of the mediums pertinent to South Asian resources and points out the important contributions of various organizations and committees in making their development possible.

INTRODUCTION

South Asia, one of the major realms of Asian continent, covers a large geographical area consisting of Bangladesh, Bhutan, India, Maldives, Nepal, Pakistan, and Sri Lanka. Because of South Asia's ancient history, civilization, variety of cultures, and complex society, American scholars are becoming more involved in teaching and research on its various aspects. Until 1920s, research studies on South Asia were concentrated mainly in Sanskrit, religions, and philosophy. The University of Pennsylvania was first to start a South Asian program, with the generous support of the Carnegie Foundation in 1948. Since then, South Asian studies programs have been established in all major universities in North America.

91

To support this continuing interest, over the years, libraries have developed a good collection of print and microform resources. Over fifty university libraries in United States have acquired and organized comprehensive collections of books, microfilms, journals, and newspapers on various aspects of South Asia through the Library of Congress's South Asia Cooperative Acquisitions Program. Under this program, the New Delhi office and its suboffices in Dhaka, Kathmandu, Colombo, and Rangoon (now Yangon) identify, selectively acquire, catalog, and distribute a wide variety of library materials from Bangladesh, India, Maldives, Mongolia, Myanmar, Nepal, and Sri Lanka, as well as Tibetan language publications. Also, the Pakistan office in Islamabad acquires a variety of materials from Pakistan, Iran, Central Asia, and Afghanistan for research libraries in United States. Monographs, serials, newspapers, and special materials in eleven languages (including Urdu, Persian, English, and others) are acquired through commercial book dealers and representatives, by exchange arrangements and through local gifts. Besides this, several monographs and serials published on South Asia in Europe, North America, and other countries are procured through regular book trade channels. Similarly, twenty-two university libraries in Canada have developed voluminous collections on South Asia through the efforts of the Shastri Indo-Canadian Institute. Books, serials and newspapers are procured by the New Delhi office and then distributed to the participating university libraries in Canada.

MICROFORM DEVELOPMENT

In 1963, a South Asian Microform Committee was constituted to acquire and maintain a readily accessible collection of unique materials in microform related to the study of South Asia. In 1967, the South Asia Microform Project (SAMP) became fully operational and has since striven to cooperate with libraries and archives worldwide in preserving unique and rare materials for South Asian studies. Materials are collected both through the filming efforts of the project and through the purchase of positive copies of materials filmed by other groups, institutions, and business organizations. So far it has filmed and acquired around 23,000 scholarly printed publications. The microfilm collection is housed and administered by the Center for Research Libraries (CRL) in Chicago.

The Microfilming of Indian Publications Project (MIPP) of the Government of India and Library of Congress is also contributing by preserving and making accessible books listed in the *National Bibliography of Indian Literature*. These are books in the twenty-six major languages of South Asia, selected by a group of Indian scholars for their central importance to humanistic understanding of India. So far several thousand books have been microfilmed under this project. The IDC publisher also preserves selected South Asian rare archival documents on microform and supplies microform copies to the academic institutions.

DEVELOPMENT OF ELECTRONIC RESOURCES

In the development of collections and enhancing library services, computer technologies have made significant contribution in the handling of information activities in the university libraries in United States and elsewhere. In recent years there has been a huge proliferation of electronic and Web resources in South Asia, and information is now accessible more easily and promptly through the Internet. Also, several searchable databases and bibliographical sources have been created in the form of compact discs. Today, the major trend is to develop more and more Web resources on all aspects of South Asian studies. However, in the South Asian countries themselves, there are certain factors that make dissemination of information slow, notably the incompatibility of systems, unreliable telecommunication media, interrupted electric supply, inadequate maintenance, and shortage of resources. Efforts are continuing in the development of infrastructure to meet the current needs to handle electronic information so that it can be shared with other countries. A number of Web-based online resources have also been developed to meet the information needs of professionals and academics in European and North American institutions.

CONTRIBUTIONS OF THE DIGITAL SOUTH ASIA LIBRARY

The Digital South Asia Library (DSAL), a project of the Center for Research Libraries, was initiated to provide digital materials for reference and research on South Asia to scholars, public officials, business leaders, and other users. This project builds upon a two-year pilot project funded by the Association of Research Libraries' Global Resources Program, with support from the Andrew W. Mellon Foundation. Participants in the Digital South Asia Library include leading American universities, the Center for Research Libraries, the South Asia Microform Project, the Committee on South Asian Libraries and Documentation (CONSALD), the Association for Asian Studies, the Library of Congress, the Asia Society, the British Library, the University of Oxford, the University of Cambridge, the Sundarayya Vignana Kendram in India, Madan Puraskar Pustakalaya in Nepal, and other institutions in South Asia.

Under the aegis of DSAL, the University of Chicago, Columbia University, and the Triangle South Asia Consortium in North Carolina are creating and disseminating electronic dictionaries in various languages. For each of the twenty-six modern literary languages of South Asia, a panel of language experts identified key dictionaries currently in print and selected at least one multilingual dictionary for each language. The best available resources have been identified, and the chosen dictionaries are being converted into digital formats. When this project is completed, these dictionaries will be freely accessible to all through the Web. Besides dictionaries, a number of books, journals, reference materials, newspapers, and pedagogical books for language instruction are also accessible through the DSAL home page (http://dsal.uchicago.edu).

THE SOUTH ASIA LIBRARY PROJECT

The Collection Development Officers Committee of the Committee on Institutional Cooperation (CIC) is sponsoring the South Asia Project as a pilot cooperative collection management project. The South Asia collections at the University of Chicago, University of Illinois, University of Iowa, University of Michigan, University of Minnesota, and University of Wisconsin–Madison aim to provide present and future generations of South Asian scholars with the resources and services they need through the coordinated development of South Asian collections and related activities. Building on the CIC library infrastructure, the CIC South Asia Library Project seeks to provide up-to-date bibliographic information on its collection and efficient delivery of resources to its member institutions. The CIC South Asia Library Project is part of a larger effort by CONSALD (see below) libraries to better coordinate the collection management and services of South Asian collections in the United States.

CENTER FOR SOUTH ASIA LIBRARIES

The Center for South Asian Libraries (CSAL) is an American overseas research center designed to facilitate scholarly research and teaching on South Asia in all academic disciplines through improved preservation of and access to the heritage of India, Pakistan, Bangladesh, Bhutan, Nepal, and Sri Lanka, as embodied in their intellectual and artistic output in all forms. It functions as a research support facility for American scholars in the region by providing infrastructures and facilities to enhance research effectiveness and the exchange of scholarly information. These aims are accomplished through current and planned activities of CSAL, operating in conjunction with several organizations and institutions in South Asia holding similar objectives. CSAL also works closely with the Council of American Overseas Research Centers' American Overseas Digital Library. South Asia scholars and librarians in the United States, Europe, Japan, and Australia have easy access and are able to consult many of CSAL's resources via the Internet. These projects include the DSAL project, now in its second phase, and the Digital Dictionaries of South Asia project (http://dsal.uchicago.edu/csal).

THE ROLE OF CONSALD

The Committee on South Asian Libraries and Documentation (CONSALD) has been created to develop, organize, and coordinate information resources and services in Canada and United States. For last several years, South Asian bibliographers and librarians have been working together to provide excellent information resources and services to South Asians (www.lib.virginia.edu/area-studies/SouthAsia/Lib/consald.html).

THE ROLE OF THE ASSOCIATION FOR ASIAN STUDIES

Located in Ann Arbor, Michigan, the Association for Asian Studies is a scholarly, nonpolitical, nonprofit professional association open to all persons interested in Asia, including South Asia.

The association produces a useful reference source, the Bibliography of Asian Studies (BAS), which is an online version of the print *Bibliography of Asian Studies* and contains records on all subjects, especially humanities and social sciences, pertaining to East, Southeast, and South Asia published worldwide from 1971 to the present. This online database for users of Western languages includes periodical articles and books on Asia and South Asia in more than 410,000 records. It is available online via library subscription (http://ets.umdl.umich.edu/b/bas/).

SOUTH ASIA RESOURCE ACCESS ON THE INTERNET (SARAI)

The South Asia Resource Access on the Internet (SARAI) is hosted by Columbia University and is one of the best sources for finding Web-based information in South Asian studies. It provides useful links to reference and bibliographical resources, e-journals, e-news, e-books, and also the International Directory of South Asian Scholars (www.columbia.edu/cu/lweb/indiv/southasia/cuvl).

PORTAL TO ASIAN INTERNET RESOURCES

The Portal to Asian Internet Resources (PAIR), based at the University of Wisconsin–Madison, is a cooperative project of the Ohio State University Libraries, the University of Minnesota Libraries, and the University of Wisconsin–Madison Libraries. It provides faculty, students, and the interested public more than six thousand professionally selected, cataloged, and annotated online resources on Asia, including South Asia. PAIR provides a searchable catalog through which people have expeditious and easy accesses to high-quality Web resources originating in Asia that are identified, evaluated, selected, and cataloged by area library specialists. It is particularly useful for providing information related to teaching and research in those academic institutions that do not have ready access to the expertise of area library specialists and collections of major research libraries (http://webcat.library.wisc.edu:3200/PAIR/about.html).

THE UNIVERSAL LIBRARY PROJECT

The Universal Library is a project conducted jointly by institutions in the United States, China, and India. Pioneered by Jaime Carbonell, Raj Reddy, Michael Shamos, Gloriana St.Clair, and Robert Thibadeau of Carnegie Mellon University, the goal of the Million Book Project (also known as the Universal Library Project) is to digitize a million books by 2005. The task will be accomplished by scanning the books and indexing their full text with OCR technology. The undertaking will create a free-to-read, searchable digital library the approximate size of the combined libraries at Carnegie Mellon University, and one much bigger than the holdings of any high school library. (The pilot Thousand Book Project has already been successfully completed and can be accessed here.) During its pilot phase the Million Book Project is being funded by the National Science Foundation. As a pilot, it has as a goal scanning and mounting 1 percent of the world's extant titles, or approximately one million books; hence its title. Scanning began in early 2004 in China and just began in the last few months in India. Approximately 20,000 titles have been scanned as of December 2004 (http://delta.ulib.org/html).

South Asian Networks

The following selected networks are very useful to all researchers:

- INFLIBNET—The Information and Library Network Centre is the product of the Indian University Grants Commission, involved in creating infrastructure for sharing information among academic and research institutions. It is a good source for the union catalog of books, serials, and thesis databases available in university libraries in India (www.inflibnet.ac.in/index.jsp).

- SAARC—The South Asian Association for Regional Cooperation provides a platform for the peoples of South Asia to work together in a spirit of friendship, trust, and understanding. Its documentation center and library, located in New Delhi, provides information services to member states (www.saarc-sec.org).
- SAN—The South Asian Network promotes the health and empowerment of people of South Asian origin living in California and fills a critical service gap in the South Asian community, which traditionally has been underserved by public interest organizations (www .southasiannetwork.org).
- SARN—The South Asia Research Network has been created to promote the production, exchange, and dissemination of basic research information in the social sciences and humanities. It provides links to electronic publications, research notes, abstracts, research centers, and conferences (http://sarn.ssrc.org).
- SASNET—The Swedish South Asian Studies Network is a national network for research, education, and information about South Asia, based at Lund University. It encourages and promotes an open and dynamic networking process, in which Swedish researchers cooperate with researchers in South Asia and globally (www.sasnet.lu.se/ sasnetf.html).
- SAWNET—The South Asian Women Network is a useful medium of communication about South Asian women. It exists entirely in the electronic medium, and its mailing list is run by a group of volunteer moderators that reaches over seven hundred women on four continents (www.umiacs.umd.edu/users/sawweb/sawnet).

Online Resources

A number of online databases have been developed on South Asia, but most of them are in the area of science and technology. There is a great need for more online resources in the areas of social sciences, humanities, and the arts. Efforts should be made to produce some qualitative websites in various South Asian languages. In view of the vast scope of the South Asian countries and the development of online resources, this chapter attempts only a sampling of useful Internet resources and networks.

- AIIEBIP—The All India Index to English Books in Print is the first electronic version of Indian Books in Print that lists thousands of books not found elsewhere, along with their prices. A Directory of Indian Publishers with complete addresses and contact details is also included (www.nisc.com/factsheets/qebip.asp).

- SAJA—The South Asian Journalists Association fosters ties among South Asian journalists in North America and improves standards of journalistic coverage of South Asia and South Asian Americans (www.saja.org).
- SALRP—The South Asian Literary Recordings Project records the voices of prominent authors from the region reading excerpts from their works. Recordings are available in Real media and MP3 formats (www.loc.gov/acq/ovop/delhi/salrp/about.html).
- SALTA—The South Asian Teachers Association is a professional organization with the mission to encourage more effective cooperation among instructors and educators of South Asian languages, linguistics, and literature in colleges and universities in North America (http://ccat.sas.upenn.edu/salta).
- SSAS—The Society for South Asian Studies supports advanced research in history, visual and material culture, ethnography, language, religion, and literature. It publishes an annual journal, *South Asian Studies* (www.britac.ac.uk/institutes/SSAS).
- Vidyanidhi is a project to construct an Indian doctoral dissertation database, established to evolve as an online resource and funded by the Ford Foundation and Microsoft (www.vidyanidhi.org.in/home/index.html).

Online Books and Serials

A number of South Asian books are available online free on the web, and can be accessed at door.library.uiuc.edu/asx/online_books.htm.

- Serials—Web-based access to online journals can be found at (door.library.uiuc.edu/asx/serials.htm#Online).
- Newspapers—There are two popular sites that provide links to all newspapers published in South Asia: door.library.uiuc.edu/asx/online_newspapers.htm, and oldsite.library.upenn.edu/vanpelt/collections/sasia/webpapers.html.

Besides these sites, the South Asia Virtual Library serves as a directory that keeps track of Web-based sources for the South Asian section and is a starting place for exploration. It can be accessed at www.columbia.edu/cu/lweb/indiv/southasia/cuvl. Also, the South Asian libraries and information centers provide links to various libraries at door.library.uiuc.edu/asx/SA_libraries.htm. A listing of American and Canadian libraries with South Asian collections is available at www.columbia.edu/cu/ lweb/indiv/southasia/cuvl/LIBS.html.

SOME SUGGESTIONS

In the future we will need to have more cooperative and coordinated efforts for delivery of information to South Asian scholars. Keeping in view our inadequate funding, shrinking space, and reduction in staff time, we have to carefully consider the scope of our acquisition policies in respect to South Asian materials. Only material that is essential for our teaching and research programs should be procured for our libraries. Librarians should consider revising profiles each year so that undesired materials may not occupy library shelves. For a better and specialized information service, there is an acute need to hire and retain South Asian librarians with language, cultural, technical, and subject skills. We do not have many bibliographies dealing with research materials related to specific topics of South Asian studies, so there is a need to develop a few standard bibliographical resources.

Along with rapid advancement in information technology, the future holds some challenges for librarians. More and more products will be available only in electronic formats. In a rapidly changing environment, with new software being developed all the time, librarians will likely encounter increased expectations from students and faculty. The result, in the coming years, will be a greater need for information literacy on the part of librarians and library users.

III

AFRICA

Chapter 7

The Role of Libraries in Combating Information Poverty in Africa

J. J. Britz and P. J. Lor

ABSTRACT

In this chapter *information poverty* is defined as characteristic of a society that lacks the required skills, abilities, or material means to obtain access to information, interpret it, and apply it appropriately. It also implies a lack of essential information and a poorly developed or culturally unfamiliar information infrastructure. Information poverty is a reality in Africa, and libraries in Africa are one of the manifestations of this form of poverty. Not only do they lack resources, but they are also low in recognition, and their survival is an ongoing struggle. Their contribution to development and education is hidden but essential. In this chapter it is argued that libraries in Africa do have an important role to play in the combating of information poverty in Africa. Three specific areas where libraries can contribute are identified, namely, access to information, the development of an information infrastructure, and development of the ability of benefit from information.

INTRODUCTION

Poverty is a concept with many dimensions. The word conjures up images of emaciated mothers and children in the Horn of Africa, of children begging on the crowded streets of a West African city, of destitute shack dwellers after yet another fire has swept through a sprawling informal settlement on the outskirts of Cape Town. Africa and poverty: This is nothing new. Week by week televised news bulletins bring these images into the homes of the affluent—at least for a minute or two, before they are replaced by stock market reports, sports stories, or advertising clips for luxury hotels, mobile phones, and skin-care products. Poverty evokes guilt, anger, and resentment. Guilt, because of the obscene contrast between the images of poverty and the comfort of the television viewer's living room. Anger, because, in a world where there are enough resource for all to live decent lives, these things simply should not be. Why aren't they doing something about this? Resentment, because our noses are being rubbed in our individual helplessness to change the world.

Poverty manifests itself in many spheres—for example, nutrition, health, employment, housing, and education. All of these are interlinked, and all of these are linked to a sphere of poverty that is a particular concern of librarians and information professionals: information poverty. In this chapter, we attempt to define and analyze information poverty, and to explore, rather tentatively, whether libraries in Africa can play a role in combating it.

INFORMATION POVERTY WITH
SPECIFIC REFERENCE TO AFRICA

In this first half of the chapter we attempt to clarify the concept of information poverty. We concentrate on its causes, three of which are highlighted: (1) lack of connectivity to information, (2) lack of availability of quality information, and (3) inability to benefit from information. The situation of Africa provides a ready reference to illustrate all three.

What Is Information Poverty?

The notion of information poverty is relatively new. As a concept it was first used in the 1950s (Britz 2004; Lievrouw and Farb 2003). The brief literature overview that follows is intended to demonstrate that information poverty is a rather complex matter and is of a political and economic as well as sociocultural nature. To be able to accommodate all the different nuances of information poverty, we propose to describe an information-poor society as a society that lacks the required skills, abilities, or material means to obtain access to information, interpret it, and apply it appropriately. There is also a lack of essential information and a poorly developed or culturally unfamiliar information infrastructure (Britz 2004, 198). Essential information is defined as that information needed to survive and develop and to be able to make critical choices—in other words, the information needed in order to live a decent life.

Intellectual capital, which represents the knowledge assets of a particular society, has the following characteristics in an information-poor society:

- The inability of people to interpret and benefit from information.
- The lack of availability and access to essential information.
- The lack of material means to pay for or gain access to information.
- The lack of skills for finding, accessing and retrieving information.
- The lack of a well-developed and well-maintained information infrastructure (Britz 2004).

Based on this description of information poverty, it is clear that there is a close relationship between information poverty and economic poverty. Economic poverty is defined as the condition in which a person is not in possession of the necessary means to live a meaningful life as well as the inability to produce such means (Lötter 2000). In both cases information plays an essential role. Without access to information pertaining to needed resources, it is impossible to satisfy basic needs. Information, and more specifically the ability to apply information, is also of cardinal importance when it comes to the ability to produce the means for living a meaningful life.

Access to information alone is in many cases not enough. People also need access to the physical products. For example, there is little to be gained from making health-care information available to a society, when there are no hospitals or medicine available to treat the sick.

WHAT CAUSES INFORMATION POVERTY?

The description of information poverty makes clear that it is a multidimensional concept. Hence a variety of factors can contribute to a situation of information poverty. We will highlight only the most important of these factors.

Lack of Connectivity to Information

Information is instrumental to all human activities; people need access to information to satisfy all their needs. To be able to survive and develop, people need to be connected to an effective information infrastructure. In today's world of advanced capitalism and globalization, modern information and communication technologies (ICTs) and, more specifically, the Internet have become synonymous with an effective information infrastructure. The development, implementation, and maintenance of a modern ICT infra-structure have also become closely associated with the so-called digital di-vide between those who have access to ICTs and those who do not. To ex-plain the lack of access to modern ICTs it must be assumed that there is a causal relationship between the material status of people and their ability to access information via ICTs.

A well-developed ICT infrastructure does not always guarantee suc-cessful use thereof. People who are poorly educated on how to use mod-ern ICTs have limited access to them and, by implication, a limited ability to benefit from digital information. Thus, although ICTs have accelerated the production and distribution of information, these technologies have at the same time exacerbated the gap between those who have access to in-formation and use it, and those who do not.

For example, the Administrative Committee on Coordination (ACC) of the United Nations (1997) views the lack of access to ICTs in developing countries as one of the primary causes of information poverty. The Or-ganisation for Economic Co-operation and Development (OECD) (2001) defines the digital divide as a gap between those who have the material means to access modern ICTs and those who do not. Doctor (1991), O'Neil (1999), and Heeks (1999) share this view. Heeks (1999, 5) remarks, "[N]ew communication technologies are revolutionising access to information—but the revolution is likely to reach everyone but the poor." According to O'Neil (1999, 3) access to ICT is limited to the so-called information elites. Buckley (1987) mentions the lack of access to computers as one of the ma-jor contributing factors to information poverty, and according to him "people without computers and access to communication lines will be the information poor in the future unless other avenues for access are pro-vided by libraries" (1987, 47).

The Social Responsibilities Discussion Group of the International Fed-eration of Library Associations and Institutions (IFLA) emphasized that the "economically disadvantaged populations of the developed countries are the information poor because, amongst other, they don't have the ma-terial means to afford modern ICTs" (Kagan 1999).

Statistics concerning access to information and the distribution of ICTs in the world confirm the ever-increasing connectivity gap between the in-formation rich and information poor. According to a World Bank report (1999), one-third of the world's population is illiterate. The percentages of illiterate persons in the population are highest in South Asia (45 percent), sub-Saharan Africa (40 percent), and the Arab states and North Africa (40 percent).

Access to ICT (more specifically the Internet) differs dramatically be-tween developed and developing countries. Recent statistics regarding access to the Internet are presented in table 7.1.

Table 7.1. Percentage of Population with Access to the Internet by Country

Continent	Country	Percentage of Population with Internet Access
Africa	South Africa	7
	Namibia	2.5
	Kenya	1.6
Asia-Pacific	China	3.6
	India	0.67
	Australia	54
	South Korea	54
Latin America	Chile	20
	Argentina	10
	Brazil	8
North America and Europe	Sweden	68
	Denmark	63
	United States of America	59
	Canada	49

Source: www.nua.com

Africa, which represents an eighth of the world's population, can surely be considered as *the* information-poorest continent when it comes to connectivity. In 1998, Africa accounted for 2 percent of the world's telephone lines. The percentage is even lower when access to and use of the Internet are considered (Mansell and Wehn 1998). This has not changed dramatically over the past five years. According to the United Nations' *2001 Human Development Report,* only 4.2 percent of the population in sub-Saharan Africa has access to a telephone, 1.1 percent use personal computers, and 0.8 percent access the Internet. However, there is some light at the end of the long, dark ICT tunnel in Africa. Africa is currently the only continent that has more cell-phone users than landline users. There are also many international initiatives to connect Africa to the digital world. These projects are definitely bearing fruit. From 2000 to 2004 the number of Internet users in Africa has increased by more that 170 percent—from 4,514,400 (2000) to 12,253,300 (2004). However, this number represents still less than 2 percent of all Internet users in the world (http://internetworldstats.com).

Lack of Availability of Quality Information

It is not only the lack of access to an effective information infrastructure that underlies information poverty. Authors such as Haywood (1995), Aguolu (1997), Norris (2001) and Burgelman and colleagues (1998) argue that the unavailability of quality information needed for development and sustainability contributes to information poverty. Schement (1995), as well as Lievrouw and Farb (2003), therefore proposes that the search for solutions for information inequalities should concentrate on information content issues.

The World Summit on the Information Society (WSIS), which was endorsed by the UN General assembly to develop and foster a political will and to take concrete steps to establish the foundations of an information society for all, focuses on the ability of all to participate in the information society and to benefit from information and knowledge sharing. The first principle proposed by the WSIS reads as follows: "A people-centered, inclusive Information Society where everyone can create, access, utilize and share information and knowledge, enabling individuals, communities and people[s] to achieve their full potential and improve their quality of life in a sustainable manner" (United Nations 2003b, 1). In the WSIS Draft Declaration of Principles it is also recognized that "technology alone cannot solve any political and social problems. ICTs should therefore be regarded as a tool and not an end in themselves" (World Summit on the Information Society 2003).

In line with the WSIS principles, IFLA strongly argues the necessity of access to quality information. According to IFLA, libraries are key players in fostering the Information Society and in bridging the so-called digital divide. Their main role is to guarantee access to information, and according to IFLA,

> [L]ibraries and information services are key actors in providing unhindered access to essential information for economic and cultural advance. In doing so, they contribute effectively to the development and maintenance of intellectual freedom, safeguarding democratic values and universal civil rights. They encourage social inclusion, by striving to serve all those in their user communities regardless of age, gender, economic or employment status, literacy or technical skills, cultural or ethnic origin, religious or political beliefs, sexual orientation and physical or mental ability. The communities they serve may be geographically based or, increasingly, linked only by technology and shared interests. (International Federation of Library Associations 2003a)

According to Aguolu (1997), access to relevant information (which he sees as a prerequisite to becoming part of the information society) will remain a myth for the developing countries until they overcome the prevailing obstacles. These obstacles include a high rate of illiteracy, unawareness of the relevance of information, poverty, and a lack of infrastructural facilities.

Burgelman and colleagues (1998) remark that the divide between information-rich and information-poor communities is *more than just digital* but also relates to the affordability, availability, and suitability of the information itself. This approach reflects the assumption that poor people do not always have the material means to access quality information. The flow of scholarly information to Africa is also very vulnerable. Rosenberg (1997), in her study on African university libraries, found that these libraries are up to 100 percent dependent on foreign donors for the acquisition of scholarly material needed for research and education. Once this funding is terminated, most of these universities will be cut off from access to much-needed scholarly information.

Schiller (1983, 1984, 1991) adds another dimension to this approach to information poverty. According to him there exists what he calls an "information gap" between those who are educated and politically and economically privileged and those at the bottom of the class system—the uneducated and the poor. In what he refers to as the *pay-per* society, the economically and educationally privileged have access to sophisticated information systems, and they have the means and skills to access and benefit from valuable information. This is specifically true of countries on the African continent. In South Africa, broadband access to the Internet costs nearly R700 ($130) per month versus a mere $15 per month, typically, in the United States. Such high cost excludes the vast majority of South Africans from regular access to the Internet. In rural sub-Saharan Africa only a few people can own a computer. A comparison between average daily wages and the cost of computers between North America and sub-Saharan Africa will explain this. In 1997 the average cost of an entry-level computer was $800 and the daily wage of a person working in North America averaged $80. In comparison, the daily wage in sub-Saharan Africa was a mere $1 a day (United Nations 2001).

Habermas, the German philosopher, can also be seen as an exponent of this approach. He expresses scepticism regarding the quality of information that is made available in the public sphere. According to him the information that is made available to the citizenry is inadequate and not reliable. It is managed and presented in such a manner that it favors only certain role players such as politicians. According to Habermas (1989), this undermines the democratic process. A prime example is Zimbabwe, where the mass media are censored and manipulated to promote the political motives of the ruling party.

The Inability to Benefit from Information

Clearly, access to quality information alone is not enough. People must have the ability to derive a benefit from information. According to Ronald Doctor (1991), we need a "right of access" in a broader sense as a "right to benefit from access" (217). This ability is determined by the individual's level of skills, experience, and other contextual factors. Information is not like food or energy where everyone requires a bare minimum to survive. Information only has value when a recipient has a need for it and the ability to process it. Otherwise information is a useless resource.

Related views include those of Akhtar and Melesse (1994), who see the problem of information poverty as an extremely complex one that encompasses factors such as attitudes, managerial skills, and finances:

> The general lack of appreciation of the role of information, the almost non-existent national information policies and the recurrent, inadequate financial resources allocated to information systems and networks development and maintenance have severely deterred the use of information to solve Africa's socio-economic problems. (314)

Fahey (2003) and Nath (2001) relate information poverty to the inability of humans to benefit from the use of information. Nath refers to this as a problem of the mind, because many developing countries have an inability to "recognise the knowledge they possess, put a value to it and use the power of knowledge to their growth" (2001).

Tapscott (1995), Ponelis (1998), Mosco (2000), and Warschauer (2003) link information poverty directly to a lack of education. Warschauer argues that we must rethink the so-called digital divide. According to him, ICT is embedded in a "complex array of factors encompassing physical, digital, human and social resources and content" (2003). He prefers to call the current divide a literacy divide, in which literacy is understood as a "set of social practices rather than a narrow cognitive skill" (2003). Tapscott (1995, 294) emphasizes the importance of education, which must, according to him, be seen as central to addressing the problem of information poverty. According to Ponelis, the information poor can be defined as those who lack information (literacy) skills, such as the ability to locate data leading to information, choose from among a variety of sources, and analyze and interpret what has been gathered for relevancy and accuracy, as well as the ability to discriminate between sources of information (1998). Mosco (2000) comments that access to information should be much more that just hardware and software. "In a deeper sense, access requires a set of capabilities, intellectual, social and cultural, from basic literacy to higher education, that are necessary to make effective use of the Information Highway" (1).

It is therefore clear that a lack of education lies at the heart of information poverty. In his conclusion on the discussion of the theories of the information society, Van Audenhove (2003) writes:

> One fundamental danger then of the rhetoric—and theory—of the information society is that it mainly focuses on the technical capacity of countries and

sees education as a facilitator in the information society. I would argue that the central element in the information society is knowledge and that technology is merely the facilitator in development. The main focus in the development effort has in recent years dangerously shifted towards the technological to the detriment of the educational. (65)

In the same line of argument, Mansell and Wehn warn us not to fall into the trap of technological determinism. According to them, technology alone will not solve the problem of developing nations (and one can read into it "of information poverty"). Providing computers and access to the Internet without adding to "physical and human capital (in other words, education) is likely to increase electricity usage rather than economic growth" (Mansell and Wehn 1998, 21).

The issue is not, however, limited to basic education, basic literacy, and information literacy. What is also required is an advanced level of knowledge skills. Human capital is not limited to reading and writing skills and mastering ICT. People must also be able to apply ICT to create new knowledge. According to Mansell and Wehn, professional skills are needed to design and adapt ICTs to new uses (1998). Used in this way ICT contributes, via human capital, to development and economic growth.

The United Nations Educational, Scientific, and Cultural Organization (UNESCO) strongly supports the idea that education can solve the problem of information poverty. In a recent document, *Education in and for the Information Society* (2003), UNESCO prefers to use and promote the notion of "knowledge societies" rather than information societies, thereby emphasizing the importance of education in the information era. The World Bank also emphasizes the importance of education in Africa. According to one of the World Bank's reports on Africa, which was published in 1999, Africa is facing a human development crisis. The World Bank views human resources as an untapped and hidden resource in Africa, and argues for a massive investment in the people of Africa (World Bank Report, 1999).

Education is not the only challenge facing Africa. An even more disturbing factor is the current brain drain of well-educated people from Africa to the developed world. Africa is on the brink of a "brain collapse." For example, it is estimated that the monetary value of the exodus of people out of Africa exceeds the value of all the development aid that African countries received from the developed world (Britz and Lor 2003, 165). Some of the horrific statistics provided by the International Organisation for Migration (IOM) and the UN's Economic Commission for Africa, are:

- Between 1960 and 1975 an estimated 27,000 highly qualified Africans left the continent.

- This number increased to 40,000 between 1975 and 1984.
- The number doubled in 1987. It then represented 30 percent of the highly skilled labor force.
- Africa lost more than 60,000 professional people between 1985 and 1990 and an estimate 20,000 every year since then.

Another study by the World Bank reported that some 70,000 highly qualified African scholars and experts leave their home countries every year in order to work abroad. Africa spends an estimated $4 billion annually on recruiting some 100,000 skilled expatriates (World Markets Research Centre 2002).

Van Audenhove (2003) correctly points out that it is very questionable whether developing countries under these circumstances will ever be able to bridge, as he puts it, the "knowledge gap" (58). Meyer, Kaplan, and Charum (2001) also comment,

> The migration of skilled persons contributes to the sharpening of inequalities, both between countries and within countries, that is such a characteristic feature of globalisation. At the same time, those very inequalities as between countries further promote and underpin the process of skill migration which responds to the growing skill wage gap as between the developed and developing world. (316)

Information Poverty: A Complex Phenomenon

It has been illustrated in the previous paragraphs that information poverty is a very complex socioeconomic and cultural phenomenon and, that being so, there is no easy solution to it. Education, specifically literacy and information literacy programs, knowledge production, availability and affordability of essential information, and the development and maintenance of an adequate information infrastructure are key elements in any attempt to ameliorate the lives of millions of information-poor people living on the African continent.

However, in the search for solutions to information poverty, two important issues should be borne in mind. First, solutions to information poverty must be directly related to its causes. This implies that there should be a clear understanding of the specific causes of information poverty in a specific context to be able to address it. Each situation of information poverty requires a unique approach. Second, one of the major causes of information poverty, namely, globalization along with advanced capitalism—which is ICT driven—is not going away and must form part of the solution. This is specifically relevant to developing countries, which are marginal players in the global economy and which have to adapt to the use of modern ICT.

LIBRARIES IN AFRICA: MORIBUND OR GERMINATING?

One possible way to address at least some of the causes of information poverty in Africa is to look at the role that libraries in Africa can play. The development of libraries in Africa has not proceeded on a smooth growth path. In a paper on Africana collection development in Africa, Alemna (1997) outlined three development periods:

- The golden years, 1948–1970
- The lean years, 1971–1986
- Hopeful but uncertain years, 1987–present. (24)

With some adjustments to the dates, which would vary somewhat from country to country, and taking into account that they undeniably overlap, these three periods could be used as a framework for a general discussion of library development in sub-Saharan Africa.

The majority of African countries achieved independence in the late 1950s and in the early 1960s. In the decades immediately preceding and following independence, significant attempts were made to establish libraries and develop library services in many African countries. Optimism and energy were generated by experienced and committed expatriate librarians, funding from colonial governments, expert advice from consultants dispatched to Africa by agencies such as the British Council, and ambitious planning exercises by UNESCO (Alemna 1992a; Havard-Williams and Marco 1991; Raseroka 1994; Rosenberg 2001; Sturges and Neill 1998). In the immediate aftermath of independence, there were high expectations that libraries would play an important part in the development of the newly independent nations (Nawe 1993). It seemed as if governments were prepared to make a significant investment in library services.

The early promise was not to be fulfilled, however. Visitors to libraries that had been featured in the international library literature as flagships of library development were shocked in subsequent decades to see tattered collections in run-down buildings. Alemna (1999) described the situation of university libraries in Ghana as getting dangerously close to being "inert and moribund":

> The materials in our university libraries are mostly outdated and of little or no academic and research value. As a result of the lack of new materials, the acquisitions and cataloguing departments have been left inactive. The poor budgetary allocations have left the acquisitions librarians almost incapacitated. (7–8)

A major report on African university libraries, based on eighteen case studies in eleven countries (Rosenberg 1997) confirmed that this was by no means atypical. A similar decline was occurring in the state of African public libraries (Mambo 1998; Raseroka 1994). A report on the state of public libraries in ten anglophone African countries

> portrays the public library movement in Africa as being very weak, with numerous problems regarding financial constraints, lack of human resources, outdated materials and poor use. The only sector in African society that uses public libraries is school children . . . [who] do not use the materials held in the library but use libraries primarily as places for study, because they are quieter and more spacious than their homes. (Issak 2000, 3)

National libraries, many of which have dual roles as repositories of national heritage and as public library services, fare no better (Lor 2000, 2003), while in many countries school libraries are in a state of decline. Only special libraries have "fared marginally better" (Aiyepeku and Komolafe 1997, 63).

During this period, sustainability became a key issue for donor agencies. Excessive reliance on donations appeared only to lead to increasing dependence. For example, universities simply stopped funding library acquisitions, expecting donors, or successions of donors, to ensure a continuing flow of material to update their collections (Rosenberg 1997). As project after project turned out not to be sustainable after its donor funding ran out, the question arose: Can libraries be sustained in Africa?

Much writing on the state of libraries in Africa constitutes a litany of woes, which can be summarized as lack of use, lack of appreciation, and lack of resources. These are interlinked and are generally blamed on a variety of problems external to the library itself, such as illiteracy, the absence of a reading culture, an education system inimical to reading, lack of books in local languages, lack of government policies supporting libraries, economic difficulties, fiscal austerity measures, inappropriate donor interventions, lack of foreign exchange, and lack of infrastructure (cf. Mchombu 1991, 186–87).

But as Mchombu perceptively pointed out, these are external problems, not root causes. The third of the development periods listed at the beginning of this section is characterized by reflection and soul-searching on the part of African librarians in an attempt to arrive at the root causes for the failure of libraries to thrive on this continent. From this process, perhaps, a truly African librarianship can emerge (Mchombu 1982, 1998; Nawe 1993; Sturges and Neill 1998).

Referring to the commonly identified problems, Mchombu (1991) suggests that these are not necessarily the *causes* of the weakness characterizing African librarianship. Instead they may be the results of the failure of African librarianship to "adapt to Africa's social, economic and cultural realities" (187). What, then, are the underlying problems or root causes? The first is the burden of the Anglo-American model that was imposed on the African situation, making the library an "alien institution":

> Since then, African libraries have found it very difficult to stoop and draw nourishment from their own people, and in turn to enrich their environment. Instead, the libraries have remained aloof and isolated and have been content to serve the minority rather than develop innovative services and form alliances which would have permitted services for both the minority elite and the majority with low levels of education. (Mchombu 1991, 188)

A second problem identified by Mchombu (1991) is the "structural decoupling of libraries in Africa from their key user target groups, and the development of an inward-looking mentality which tends to glorify internal processes at the expense of maximising use of library resources" (188). He proposes a development-oriented restructuring of libraries with the emphasis on supplying information rather than guarding documents. Librarians should "assume the role of flexible facilitators of information transfer and exchange within institutions, groups and society at large" (189).

Reliance on the information and knowledge base of the developed countries is a related problem. Mchombu (1991, 189) warns against shortsighted reliance on book donations, which can be an instrument of "intellectual and mental subjugation." Cram (1993) argues that "[l]ibrary colonialism— the domination of Africans by Western nations through the use of information power—remains one of the most hidden but deadly instruments of neo-colonialism" (13–14).

The rethinking of librarianship in Africa is proceeding in two overlapping phases: first, adapting Western librarianship to Africa; second, realizing that a more radical approach is necessary (Rosenberg 2001; Sturges and Neill 1998). Adaptation can entail quite simple, practical changes. For example, in South Africa the public library branches in predominantly white suburbs provided very little seating. Users came into the library mainly to return borrowed books and take out new ones. Not many stayed to sit and read. Today, the design of new public library branches has to make much greater provision for reader places, to accommodate large numbers of students who want to spend all day in the library. In a similar vein, one may expand the Dewey Decimal Classification to make better provision for African languages and literatures, religions, and philosophy, or add African subject headings to the Library of Congress subject headings. This leaves the library as an institution untouched. Taking adaptation a step further, librarians are adding books in African languages (insofar as they can be found) to their collections, along with easy reading materials for newly literate adults and audiovisual media for the use of illiterate persons. A greater awareness of the needs of the illiterate, people living in informal settlements, rural people, people with disabilities, and people affected by HIV/AIDS, for example, accompanied by programs to reach out to them with appropriate materials and modes of delivery, bit by bit eats into the hard shell of the Eurocentric public library and begins to blur its Anglo-American profile. But the basic institution remains.

Sturges and Neill (1998) suggest that such measures are little more than "sticking plaster on wounds that require surgery" (136). They argue that a new paradigm of information is required, based on six principles: financial realism, self-reliance, sustainability, democracy, responsiveness, and communication. Rosenberg (2001) identifies three fundamental principles for redefining the role and purpose of libraries in Africa:

- Along with other agencies, libraries are a means of information transfer. They will only survive if they meet an unmet need in the information transfer process.
- The real information needs of users must be determined and addressed, even if the resulting agency does not conform to the Anglo-American model.
- Africa is a poor continent and any information service must be cost-effective. Planning must take financial realities into account.

The first point implies that libraries, however well established they might be in developed countries, do not have a divine right of existence. To survive in Africa they have to play a functional role in information transfer.

The second point implies that the languages, content, and modes of delivery of collections and services must be relevant and accessible to Africans and African culture in a world dominated by Western science and communication media. The library as an "alien implant" (Sturges and Neill 1998), based on the Anglo-American model, is not adequate in meeting the needs of Africa.

The third point was first raised by Kingo Mchombu (1982) in a seminal article entitled "On the Librarianship of Poverty." In it he stated that "the chief factor determining information work in developing countries should be poverty rather than affluence" (242). Poverty is the root cause of such factors as lack of trained staff and a weak publishing industry. It is no use ignoring this. Therefore, standards of library provision drawn up in developed countries cannot be applied in the developing world without causing grave distortion. For example, if library buildings are erected to exacting Western standards, it may take many decades, or even centuries, before the library system can reach the whole country. Mchombu (1982) proposed that

> the standards of information services must be tailored to the economic ability of a country. If the pattern of information services is pushed ahead of general economic development, standards will be set that can only be maintained in small pockets of the country. The lucky few may have very good service, but most people will have no service at all, or a service that is inadequate and at prohibitive distances. (245)

Thus libraries in Africa have to exist in a situation of poverty. They are not outside this situation. They themselves are poor and a manifestation of the information poverty under discussion in this chapter. If this basic fact is not faced, no amount of talking or theorizing is going to make any difference. The key requirement is sustainability, which implies that the agencies, institutions,or mechanisms that have been set up must be able to survive without indefinite donor funding, in a context of poverty, where government funding cannot be relied on and resources are hard to come by.

In the context of information poverty, where does this leave us? For the purpose of this chapter, two themes stand out: relevance and poverty. They form the background to the final section.

LIBRARIES AS MANIFESTATIONS AND ADVERSARIES OF POVERTY

The conditions of libraries in Africa are a manifestation of information poverty. Yet libraries also have a role to play in combating information poverty. We now briefly consider the role of libraries in addressing the three causes of information poverty: lack of connectivity to information, unavailability of quality information, and inability to benefit from information.

Lack of Connectivity to Information

Connectivity is a major problem for libraries in Africa. A great deal of scientific information is now published on the Web. Although the costs of access set by scientific, technical, and medical publishers can put electronic journals beyond the reach of African scholars, there are various schemes to provide scholars in the poorer countries with affordable access, for example, the Health InterNetwork Access to Research Initiative (HINARI), Global Online Research in Agriculture (GORA), and the INASP Programme for the Enhancement of Research Information (PERI)

(Lor and Britz forthcoming). Open-access journals and institutional repositories also make more and more scholarly materials available to African scholars free of charge. However, to access these materials, African scholars need bandwidth. In a recent discussion of information and communications technologies (ICTs) in African libraries, an American delegate remarked that he had more bandwidth available to his home than a major African university represented at the meeting. There is a huge connectivity gap between African scholars and their colleagues in developed countries. At the same meeting delegates learned that, for reasons best known to the governments and telecommunications corporations controlling this facility, bandwidth provided by the new submarine cable encircling Africa is not being made available to universities and public libraries in West Africa (Lor 2004b). Unfortunately, librarians are far removed from the boardrooms where decisions are made on the availability of ICT infrastructure. But they could play a role by mobilizing their client communities and providing them with the information they need as ammunition in the battle for affordable access.

Lack of Availability of Quality Information

More important than the ICT infrastructure is the content that is conveyed by it. ICTs are a tool, not an end in themselves. It is generally recognized that up-to-date information of good quality is essential for development. If we bridge the digital divide, in which direction will the information flow? At first glance the answer is plain: The enormous information resources of the developed countries should be brought to bear on the development challenges of Africa, on the killer diseases that decimate African populations, on the pests that consume African crops, and on problems of democratic governance, social services, and economic management. Above all, such information is needed for education and research. Africa can never be self-sufficient in scientific and technical knowledge. Her production of books and research articles is a fraction of that of the world as a whole (Lor 1996). There is a moral obligation on the developed countries to share their wealth of knowledge with the developing countries (Britz 2004). This knowledge must be disseminated widely, not merely to local subsidiaries of multinationals or to small elite groups, to improve production and generate more profits for shareholders in North, but to support education at all levels. The new generation must be empowered to exploit all the knowledge and technologies that are the common resources of humankind, insofar as these can be adapted and utilized in Africa to ensure a better quality of life for all.

Libraries in Africa should play important roles in serving as conduits for knowledge to flow from North to South. Universities and research institutes need to offer excellent facilities to retain the kinds of well-educated scholars and professionals who are now leaving Africa by the thousands. Such facilities would necessarily include access to the world's scholarly literature. Libraries should also provide access to knowledge for the empowerment of the general population. Libraries are agencies that in principle are open to all, and they do not have a hidden agenda to sell things to people. All should be able to enter and gain access to knowledge, without let or hindrance.

Apart from open access to online materials, there are three basic mechanisms by which African libraries can acquire information resources from developed countries. The obvious mechanism is purchase, but, as we have already shown, generally African libraries do not have the funds to purchase more than a small fraction of what their users need. A second mechanism is exchange of publications. African institutions may have little to exchange, but many institutions in developed countries do not raise objections if there is a severe imbalance in the value of the material exchanged. In any case, the African material, even if of little commercial value, would be difficult to acquire by other means. The problem is that maintaining exchange relationships is very labor-intensive. This is not so much a problem in Africa as it is in the developed countries, where labor is expensive. A third mechanism of acquisition is by gift. Since World War II millions of books have been donated to African libraries (Rosen 2001; Watkins 1997; cf. Zell 1987). There are a good many organizations in the developed countries that run book donation programs (Watkins 1997; Zell 1987). Most programs are altruistically motivated (cf. Rosen 2001; Watkins 1997), but this is not true in all cases. Sturges and Neill (1998) cite examples of entirely inappropriate donations to African libraries. In some cases the real beneficiaries are not African students but American publishers, who earn handsome tax cuts by making these donations (cf. Rosen 2001). Even well-intentioned programs without ulterior motives may have unintended negative consequences. African publishers, for example, are concerned about "book aid" that might flood their limited markets (cf. Mutula and Nakitto 2002; Zell 1992) and wipe out their fragile industry.

The value of whatever useful materials are acquired should be maximized by excellent organization and management of collections and by resource sharing between the libraries of developing countries. The latter principle is more honored in the breach than in the observance, unfortunately. It would seem that, paradoxically, the wealthiest countries have the best resource sharing systems, while poverty inhibits sharing. In Africa resource-sharing is poorly developed (Rosenberg 1993).

We posed the question, If we bridge the digital divide, in which direction will the information flow? Not in one direction only: it should also flow from Africa to the developed countries. Africa has much to offer the world. Apart from African contributions to science and scholarship, which are unjustifiably neglected (Britz and Lor 2003), Africa has a wealth of oral traditions and indigenous knowledge, conveyed in a vast number of languages. In sub-Saharan Africa the Niger-Congo group of languages, which is by far the largest, but not the only, language group on the continent, includes an estimated 1,400 languages and many more dialects ("Niger-Congo" 2004). If every language embodies a unique way of seeing the world, Africa cannot really be called information-poor.

Perhaps the problem in Africa is not the lack of knowledge but the undervaluation thereof, due to the influence of the colonial powers and postcolonial scientific and cultural dominance. This leads Africans to discard indigenous knowledge and culture in favor of "scientific" knowledge and to look down on traditions as "backward":

> In most African countries literacy and education developed as a result of the European advent—and in consequence the people were drawn into an alien culture which disrupted their own. The literature of the earlier missionaries seemed to tell them to despise their own way of life; on the other hand, the literature of the modern scientific world led them to ignore it as irrelevant. (Benge 1970, 186)

The prevalence of such attitudes suggests that information poverty is also in the mind. Mchombu (1991) warns against the uncritical acceptance of "foreign knowledge and information, some of which distorts their own reality" (189). As suggested by Cram (1993), "A true African library would be one that Africans and others could walk into to experience the realities of the African world view" (15).

It is therefore an important task of African libraries to add value to Africa's own heritage—its indigenous languages, cultural traditions (folk tales, praise poetry, songs, dances, etc.) and its indigenous knowledge— by collecting, preserving, and organizing it and making it available, all this being done with great sensitivity to its cultural context. A further task is to promote awareness of this heritage and to interpret it for library users so that it can be appreciated on its own terms, not as a primitive curiosity to be gaped at (Alemna, 1999, 1992b; Lor forthcoming). Oral tradition and indigenous knowledge should be used as the basis for books in indigenous languages, which make mother tongue instruction possible. The community speaking the language should be involved in the production of teaching materials for their schools and adult literacy projects. It is not only the recording of languages that can help to prevent their extinction. Their recognition and use in a high-status activity such as education helps to preserve their status (Lor 2004a). "This is critical to the self-esteem of the community: Any perceived diminution in the social standing of a language will undermine its members' sense of identity, and . . . the total loss of a language is likely to shatter it completely" (Abbott 2002, 227).

According to Mchombu, African countries need to reduce their information dependency on Europe and the United States. This does not mean that they should isolate themselves from the rest of the world but that they should "create their own knowledge and information base," which is more relevant to the African context and can incorporate oral tradition and indigenous knowledge. African librarians must bridge the divide between Africa's traditional knowledge resources and Western knowledge resources (Mchombu 1991). If information relations are seen not only in North-South terms but also in terms of South-North and South-South information flows, information poverty becomes a more multidimensional and relative concept, and possibly less intimidating.

The Inability to Benefit from Information

To benefit from information, people must be aware of needing it and they must be aware of its potential value to them. Poor and illiterate people may be unaware of having "information needs" as such. Information is an abstract concept that does not seem relevant to their daily struggle for survival. However, many studies have shown that people in all sectors of society have information needs and that the provision of information services can make a difference to their lives (cf. Mchombu 2001). But librarians should not sit back passively and wait for the people to come and ask for information.

It is important, therefore, that libraries in Africa take up the challenge of educating their clients and potential clients, particularly those at the grassroots level, who are at risk of exploitation, and of reaching out to them. Aboyade (1989) identified five roles for libraries: (1) reinforcement of specific messages already disseminated by other agencies (such as community development and agricultural extension agencies); (2) repackaging of information; (3) acquiring and organizing specialized materials (which would presumably include literacy materials and reading materials for newly literate adults); (4) filling identified information needs; and (5) coordinating information transfer activities among the people.

Illiteracy is a major challenge to African libraries, but direct involvement in literacy teaching, as distinct from information-literacy teaching, does not appear to be common. This may reflect a reluctance of librarians to take on the role of another professional group (teachers), as well as their fixation on collections and correct processing thereof, as already referred to. Some diffidence may not be entirely out of place. Cram (1993) points out that literacy has a negative side: "The essentially negative side of literacy is the way it dispossesses speech by giving the illiterate the impression that books are the only possible vector of culture and by teaching him to devalue his own discourse in his own eyes" (17).

This is all the more reason for librarians to be involved. Ideally librarians should not only provide venues for literacy classes and literacy materials for the students, but they should also act as mentors to literacy students and the newly literate, guiding them to appropriate reading matter in their own languages and contextualizing their reading within their society and within the diversity of the world's culture and knowledge systems. This is a high ideal, unlikely to be widely attained without extensive reeducation of librarians. At least librarians are more comfortable with the teaching of information literacy, which—with the coming of the Web and widespread naïve reliance on Google—is more critical than ever before to ensure that library users are able to benefit from the information that is made accessible to them.

CONCLUSION

Poverty is real. Information poverty is a fact of life in Africa. Libraries themselves are a manifestation of information poverty. Lacking recognition, low on the development agenda, constrained by lack of resources, constantly getting the tail end of budgets, Africa's libraries are engaged in an ongoing struggle for survival. Because they play a supporting role, it is difficult to quantify the contribution libraries make to development. One may still quantify the contribution of higher education: So many teachers, vets, or agronomists are produced in a given period. Hidden in the production figure is the contribution the library has made to the training of these graduates. Even more invisible is the contribution the university or college library may have made to the quality of their education and the contribution that national library services can make to the quality of their continuing professional practice. Nevertheless, libraries do have a role to play in combating information poverty—not only poverty measured in lack of access to the information resources of the developed countries but also the poverty in the mind that results from failure to cherish Africa's own information wealth.

BIBLIOGRAPHY

Abbott, G. 2002. "The Importance of Activating Indigenous Languages in the Drive for Development." *Information Development* 18, no. 4: 227–30.

Aboyade, B. 1989. "Communication and Information in Contemporary African Society: The Role of the Library." Pp. 1–3 in *Proceedings of the 55th IFLA General Conference, Paris*. Munich, Germany: K. G. Sauer.

Aguolu, I. E. 1997. "Accessibility of Information: A Myth for Developing Countries?" *New Library World* 98: 25–29.

Aiyepeku, W. O., and R. S. Komolafe. 1997. "Africa." Pp. 62–71 in *World Information Report 1997–98.* Paris: UNESCO.

Akhtar, S., and M. Melesse. 1994. "Africa, Information and Development: IDRC's Experience." *Journal of Information Science* 20, no. 5: 314–22.

Alemna, A. A. 1999. "Libraries in Research and Scholarship in Ghana." *African Research and Documentation* 57: 6–14.

———. 1992a. "The Role of External Aid in Ghana's Library Development." *Library Review* 41, no. 5: 32–41.

———. 1992b. "Towards a New Emphasis on Oral Tradition as an Information Source in African Libraries." *Journal of Documentation* 48, no. 4: 422–29.

———. 1997. " Collection Development of Africana Materials." In *Africana Librarianship in the 21st Century: Treasuring the Past and Building the Future,* 40th Anniversary Conference of the Africana Librarians Council. Columbus: Ohio University (unpublished).

Audenhove, L. Van. 2003. "Theories of the Information Society and Development: Recent Theoretical Contributions and Their Relevance to the Developing World." *Communication* 29, nos. 1 and 2: 48–67.

Benge, R. C. 1970. *Libraries and Cultural Change.* London: Bingley.

Britz, J. J. 2004. "To Know or Not to Know: A Moral Reflection on Information Poverty." *Journal of Information Science* 30, no. 3: 192–204.

Britz, J. J., and P. Lor. 2003. "A Moral Reflection on the Information Flow from South to North: An African Perspective." *Libri* 53, no. 3: 160–73.

Buckley, F. J. 1987. "Knowledge—Access Issues." *The Information Society* 5, no. 1: 45–50.

Burgelman, J. C., et al. 1998. "De Geschiedenis Herhaalt Zich . . . Altijd Anders. Noordzuid Cahier. De Digitale Kloof. De Informatierevolutie en Het Zuiden." *Noordzuid Cahier* 24, no. 4: 11–24.

Cram, J. 1993. "Colonialism and Libraries in Third World Africa." *Australian Library Journal* 42, no. 1: 13–20.

Cronin, B. 1995. "Social Development and the Role of Information." *The New Review of Information and Library Research* 1:23–37.

Doctor, R. D. 1991. "Information Technologies and Social Equity: Confronting the Revolution." *Journal of the American Society for Information Science* 42, no. 3: 216–28.

Fahey, N. 2003. "Addressing Information Poverty: An Australian Experience." http://acqol.deakin.edu.au/Conferences/paperFahey1.PDF (accessed January 12, 2004).

Habermas, J. 1989. *The Structural Transformation of the Public Sphere,* translated by T. Burger. Cambridge, MA: MIT Press.

Havard-Williams, P., and G. A. Marco. 1991. "Time, Development, Africa." *Alexandria* 3, no. 2: 81–88.

Haywood, T. 1995. *Info-rich—Info-poor. Access and Exchange in the Global Information Society.* London: Bowker-Saur.

Heeks, R. 1999. "Information and Communication Technologies, Poverty and Development." Pp.1–19 in *Development Informatics. Working Paper Series, No. 5.* Manchester, UK: Institute for Development Policy and Management.

International Federation of Library Associations (IFLA). 2003a. *Document WSIS.* www.ifla.org (accessed June 15, 2004).

———. 2003b. *IFLA Report for a Round-Table Meeting, March 2003, Lugano.* www.ifla .org/III/wsis/wsis-lugano.pdf (accessed January 12, 2004).

Internet Usage Statistics—The Big Picture. World Internet Usage and Population Statistics. 2004. www.internet worldstats.com/stats.htm (accessed July 24, 2004).

Issak, A. 2000. *Public Libraries in Africa: A Report and Annotated Bibliography.* Oxford: International Network for the Availability of Scientific Publications.

Kagan, A. 1999. "The Growing Gap between the Information Rich and the Information Poor, Both within Countries and between Countries. A Composite Policy Paper of the Social Responsibilities Discussion Group, International Federation of Library Associations and Institutions." www.ifla.org/VII/dg/srdg/srdg7.htm (accessed June 2001).

Kularatne, E. D. T. 1997. "Information Needs and Information Provision in Developing Countries." *Information for Development* 13, no. 3: 117–21.

Lievrouw, L. A., and S. E. Farb. 2003. "Information and Equity." *Annual Review of Information Science and Technology,* 37: 499–538.

Lor, P. J. 1996. "Information Dependence in Southern Africa: Global and Subregional Perspectives." *African Journal of Library, Archives and Information Science* 6, no. 1: 1–10.

———. 2000. "Libraries in the African Renaissance: African Experience and Prospects for Survival in the Information Age." *International Information & Library Review* 32, no. 2: 213–36.

———. 2003. "What Prospects for National Libraries in Africa? A South African Perspective." *Alexandria* 5, no. 3: 141–50.

———. 2004a. "Indigenous Knowledge, Minority Languages and Life-Long Learning: A Role for Public Libraries." Pp. 59–70 in *Libraries for Lifelong Literacy: Unrestricted Access to Information as a Basis for Lifelong Learning and Empowerment,* edited by S. Seidelin, S. Hamilton, and P. Sturges. Copenhagen, Denmark: IFLA/FAIFE.

———. 2004b. "Report on the Workshop on ICTs and the Library: Experiences, Opportunities and Challenges for Libraries in Africa, Benoni, South Africa, 20–23 July 2004." www.inasp.info/lsp/ict-workshop-2004/lor-report.doc (accessed September 17, 2004).

———. forthcoming. "Storehouses of Knowledge? The Role of Libraries in Preserving and Promoting Indigenous Knowledge." *Indilinga: African Journal of Indigenous Knowledge Systems.*

Lor, P. J., and J. J. Britz. forthcoming. *Knowledge Production, International Information Flows and Intellectual Property: An African Perspective.*

Lötter, H. P. P. 2000. *Christians and Poverty.* DD Thesis. Pretoria: University of Pretoria, South Africa. www.nua.com (accessed September 22, 2003).

Mambo, H .L. 1998. "Public Libraries in Africa: A Critical Assessment." *African Journal of Library, Archives and Information Science* 8, no. 2: 67–76.

Mansell, R., and U. Wehn, eds. 1998.*Knowledge Societies: Information Technology for Sustainable Development.* Oxford, UK: Oxford University Press (for the United Nations Commission on Science and Technology for Development).

Mchombu, K. 1982. "On the Librarianship of Poverty." *Libri* 32, no. 3: 241–50.

———. 1991. "Which Way African Librarianship?" *International Library Review* 23: 183–200.

———. 1998. "African Librarianship: Reality or Illusion?" Pp. 150–56 in *Libraries: Global Reach, Local Touch*, edited by K. de la Pena McCook, B. J. Ford, and K. Lippincott. Chicago: American Library Association.

———. 2001. "Research on Measuring the Impact of Information on Rural Development." Pp. 229–38 in *Knowledge, Information and Development: an African Perspective*, edited by C. Stillwell, A. Leach, and S. Burton. Pietermaritzburg, South Africa: School of Human and Social Studies.

Meyer, J. B., D. Kaplan, and J. Charum. 2001. "Scientific Nomadism and the New Geopolitics of Knowledge." *International Social Science Journal* 168: 309–21.

Mosco, V. 2000. "Public Policy and the Information Highway: Access Equity and Universality. A Report to the National Library of Canada," contract number 70071–9-5107. www.carleton.edu/~vmosco/pubpol.htm (accessed April 24, 2001).

Mutula, S. M., and M. M. T. Nakitto. 2002. "Book Publishing Patterns in Uganda: Challenges and Prospects." *African Journal of Library, Archives and Information Science* 12, no. 2: 177–88.

Nath, V. 2001. "Heralding ICT Enabled Knowledge Societies. Way Forward for the Developing Countries." www.cddc. vt.edu /knownet/articles/heralding.htm (accessed September 2, 2002.)

Nawe, J. 1993. "The Realities of Adaptation of Western Librarianship to the African Situation." *African Journal of Library, Archives and Information Science* 3, no. 1: 1–9.

"Niger-Congo Languages." n. d. *Encyclopaedia Britannica online*. http://search.eb .com/eb/article?eu=118153 (accessed June 18, 2004).

Norris, P. 2001. *Digital Divide: Civic Engagement, Information Poverty, and the Internet Worldwide*. Cambridge: Cambridge University Press.

O'Neil, D. V. 1999. "Ubiquitous Access to Telecommunication Technologies: Is Access a Positive Freedom? Intellectual Property and Technology." http://infoeagle .bc.edu/bc_org/avp/law/st_org/iptf/commentary/content/1999060402.html (accessed November 30, 2000).

Organisation for Economic Co-ordination and Development (OECD). 2001. *Understanding the Digital Divide*. Paris: OECD. www.monash.edu.au/casestudies/ css/267_dd.htm (accessed May 16, 2002).

Ponelis, S. R. 1998. *Information Wealth and Information Poverty*. Pretoria: University of Pretoria (unpublished).

Raseroka, H. K. 1994. "Changes in Public Libraries during the Last Twenty Years: An African Perspective." *Libri* 44, no. 2: 153–63.

Rosen, J. 2001. "IBB, Sabre Ship Thousands of Books Overseas." *Publishers Weekly* 248, no. 2: 20.

Rosenberg, D. 1993. "Resource Sharing—Is It the Answer for Africa?" *African Journal of Library, Archives and Information Science* 3, no. 2: 107–12

———. 1997. *UniversityLibraries in Africa: A Review of Their Current State and Future Potential*. 2 vols. London: International Africa Institute.

———. 2001. "The Sustainability of Libraries and Resource Centres in Africa." Pp. 11–24 in *Knowledge, Information and Development: An African Perspective*, edited by C. Stillwell, A. Leach, and S. Burton. Pietermaritzburg, South Africa: School of Human and Social Studies.

Schement, J. R. 1995. "Beyond Universal Service: Characteristics of Americans without Telephones, 1980–1993." *Telecommunications Policy* 19: 477–85.

Schiller, H. I. 1983. "The Communications Revolution: Who Benefits?" *Media Development* 4: 8–20.

———. 1984. *Information and the Crisis Economy.* Norwood, NJ: Ablex.

———. 1991. "Public Information Goes Corporate." *Library Journal* 51, no. 8: 42–45.

Sturges, R. P., and R. Neill. 1998. *The Quiet Struggle: Information and Libraries for the People of Africa.* 2nd ed. London: Mansell.

Tapscott, D. 1995. *The Digital Economy: Promise and Peril in the Age of Networked Intelligence.* New York: McGraw-Hill.

UNESCO. 2003. *Education in and for the Information Society.* Geneva: United Nations.

United Nations. 2001. *Human Development Report.* New York: Oxford University Press

———. 2003b. *Libraries and the Information Society.* Geneva: United Nations.

———. 1997. Administrative Committee on Co-ordination. *Statement on Universal Access to Basic Communication and Information Services.* New York: United Nations.

Van Audenhove, L. 2003. "Theories of the Information Society and Development: Recent Theoretical Contributions and Their Revelance to the Developing World." *Communication* 29 (H2): 48–71.

Warschauer, M. 2003. "Reconceptualising the Digital Divide. www.firstmonday .dk/issues/issue7_7/warschauer/ (accessed December 1, 2003).

Watkins, C. 1997. "Changing the World through Books." *American Libraries* 28, no. 9: 52–54.

World Bank. 1999. *World Development Report. Knowledge for Development.* New York: Oxford University Press.

World Markets Research Centre. 2002. "The Brain Drain—Africa's Achilles Heel." www.worldmarketanalysis.com/InFocus2002/article/africa_braindrain.html (accessed May 22, 2004).

World Summit on the Information Society. 2003. "Summit Draft Declaration of Principles, 2003." www.itu.int/dms_pub/itu-s/md/03/wsispc3/td/030915/ S03-WSISPC3-030915-TD-GEN-0001!R2B!MSW-E.doc (accessed February 14, 2004).

Zell, H. M. 1987. "The Other Famine." *Libri* 37, no. 4: 294–306.

———. 1992. "Africa: The Neglected Continent." Pp. 65–76 in *Publishing and Development in the Third World,* edited by Philip G. Altbach. London: Hans Zell Publishers.

Chapter 8

The Impact of Information Communications Technology on Academic Libraries in Sub-Saharan Africa, with Specific Reference to Botswana

H. Kay Raseroka

ABSTRACT

This chapter reviews the challenges that universities and libraries in sub-Saharan Africa face. Against this backdrop it traces the response of academic libraries to the availability of information communications technologies (ICTs) suitable for data management and provision of effective information services to academic communities, in support of the core activities of learning, teaching, and research.

Factors that have led to the gradual introduction and use of ICTs are highlighted. The analysis of impact is limited to observable change in technical operations.

INTRODUCTION

Academic libraries operate within the context provided by their respective universities. The state of universities in sub-Saharan Africa has been affected by "declining resources during a period of growing enrolments, without a capacity to keep the two in balance" (World Bank 1997). Although universities in sub-Saharan Africa have been advised on the 5 percent minimum recommended allocation of expenditure for library services as a percentage of the total university expenditure, very few universities have been able to provide this minimum consistently. In the West African country of Ghana, for example, it is estimated that the library budget may

127

be as low as 2 percent. Hence, declining resources have resulted in a general educational crisis, which extends to the condition of academic libraries in general and specifically to the provision of learning support materials—from those available on print to electronic learning sources that require access capability.

For the purposes of this chapter, information communication technologies (ICTs) include computerization of library networks, application of information communication technologies for research, and access to academic information sources through exploitation of electronic resources on the Internet, digitized and print resources, digital libraries, and online information retrieval, and access to databases of bibliographic utilities.

This chapter will present the general conditions that have influenced tertiary education in sub-Saharan Africa and discuss its effects on academic libraries. It will present an overview of efforts made to install information technology infrastructure and trace the impact of ICT applications for the delivery of information services, in sub-Saharan Africa generally and at the University of Botswana, specifically.

CONTEXT OF ACADEMIC LIBRARY DEVELOPMENT

The last three decades of the twentieth century have seen a decline in national capacities to finance tertiary education, which is predominantly government owned and funded, in sub-Saharan Africa (Blair 1991; Coombe 1991). Thus annual subventions from governments to national public universities have, generally, been reduced, while the tertiary institutions are required to increase the annual student intake. This trend continues to date. Reduction of public expenditure per tertiary student in sub-Saharan Africa dropped from $3,084 in 1980 to $1,971 in 1990, while enrollment rose from 419,000 in 1980 to 1,220,000 in 1990. In the beginning of the twenty-first century the increase in student intake in public national universities is on the increase, with minimal concomitant growth of the physical and technological infrastructure. Thus academic libraries are experiencing increased demand for space and learning support facilities, including technology, to meet the increased numbers of students and teaching staff, without comparative subventions to sustain information service infrastructures.

The introduction of technology in information management has thus been perceived as a strategy for sustaining effectiveness through accessing information resources held by libraries globally and, supposedly, at reduced costs. This has been enabled by the utilization of electronic infrastructure to access needed journal articles through Internet-linked electronic networks, without necessarily owning the relevant printed journal title or collections. This strategy is a means of enhancing educational effectiveness, which is defined as the ability to find information and to be able to adapt and use it for new applications and create new knowledge. Competition for financial resources is an underlying factor for determining the allocation of scarce resources at national levels among competing demands that range from poverty alleviation to health-care delivery systems. Within academic institutions, the distribution of scarce resources generates competition between libraries, general ICT infrastructure, and laboratories, the latter being of higher standing than books.

Without the provision of facts on benefits derived from investment in ICT for information delivery to society in general and in academic institutions specifically, librarians were hard put from the 1980s to the end of the twentieth century to convince ministries of finance to allocate needed funds, on the basis of "public good" arguments. Their advocacy skills proved inadequate to promote the beneficiaries of library services as a constituency, so that university leadership would be persuaded to support libraries adequately. Hence the librarians' inadequate skills for marshalling convincing financial and advocacy arguments to justify investment in libraries in general and ICTs for information delivery specifically is perceived as one of the major causes for the poor investment in academic libraries in sub-Saharan Africa.

EXPLOITATION OF PERSONAL COMPUTERS

Recognizing the impact of ICTs on technical and information service delivery and access within the developed world in the late 1980s, academic libraries in sub-Saharan Africa invested in personal computers (PCs) as an entry into the world of technological exploitation. The magnitude and degree of ICT use, of course, varied within countries based on skilled project proposal making and allocation of financial resources available to libraries through parent institutions. This strategy made it possible for the introduction of word-processing and e-mail facilities, to a limited extent. Access points were few, but they facilitated an introduction to concepts and practical skills for electronic communication. They were means for providing skills development, understanding, and acceptance of the technology as a useful instrument for more communication and efficient management of library routines and they fostered communication, especially in accessing information among libraries as part of resource sharing.

The impact of ICT was felt at the most basic office management level: The change from centralized typing pool services to devolved word processing by all levels of staff became the order of the day. Word-processing skills became an essential skill for all library staff, as a basis for changing routine administrative functions. Yet mastery of keyboarding and word-processing fundamentals has been left largely to self-training. The lack of investment in skills development for the word-processing activity is an underlying factor for poor skills in the basic use of ICT. This deprives users of the basic advantage of technology: time-saving applications to routine tasks. Nonetheless, the introduction of ICT during the period under discussion demonstrated its effectiveness as an information transfer tool at personal, educational, and research information levels, for both library staff and users. Thus the impact of technology began to be felt, particularly, through student enthusiasm and demand for more access points to technology. Library staffs were challenged to develop means of facilitating access within limited resources, while advocating for universities to embrace ICT at the policy level, and financially.

CDS-ISIS DATABASE DEVELOPMENT

As personal computers (PCs) became more familiar to library personnel, they were used to set up small-scale automation projects to develop partial databases such as serial lists and reserve collections, with the help of freely available UNESCO CDS ISIS software. The CDS/ISIS software was, for example, used by the universities of Nairobi, Zimbabwe, and Botswana during the early 1990s to compile a list of periodicals held in all libraries in Kenya, a regional list of periodicals in Zimbabwe, and a subject databases such as gender and agriculture databases in Botswana and Malawi.

CD-ROM DATABASE EXPLOITATION

Fifteen academic libraries from southern and eastern African regions who shared ideas on survival strategies for academic libraries in the early 1990s relied heavily on the CD ROM technology as a tool for delivery of the latest information to support library users. Most of them, however, were dependent on donor funding for the purchase of their PCs and subscriptions to databases, which were generally accessible through stand-alone personal computers rather than on networked stations (Patriokios and Levey 1994). Indeed, at least six university libraries in southern Africa were pilot sites for the AAAS CD-ROM project. They benefited from a program that trained participants (librarians, students, and faculty) in evaluation and selection of subject-based CD-ROM databases, which met the academic needs of specific programs. CD-ROM databases were substituted for print journal titles that universities could no longer afford to subscribe to. The most cited CD-ROM titles used by sub-Saharan Africa academic libraries at this time for access to current research are ERIC MEDLINE, Science Citation Index, AGRIS, CAB Abstracts TROPAG, RURAL, CIMMYT Maize Germplasm Bank Enquiries System, and Biological and Agriculture Index. With the experience of CD-ROM capabilities, librarians developed in-house products that enhanced resource sharing. For example, in Kenya a CD-ROM–based compilation of theses and dissertations held by universities and research organizations is an outcome of a project whose objectives are preservation and conservation of rare local material, using microfilming as a packaging tool. The availability of suitable and affordable access to various journal titles through CD-ROM technology introduced also a possibility for the use of websites for wider information resource access (Kisenyi 2003). A few projects have been developed through collaborative efforts and in-kind commitments of university libraries, with minimal financial support from donor funding.

Although this activity started in the late 1980s, the situation is still prevalent in various academic libraries in a number of African countries that have been experiencing unstable political environments during the 1980s and 1990s. For example, academic libraries in the Democratic Republic of Congo (DRC) do not have significant penetration of ICTs. There are only privately funded efforts to provide medical students in Kinshasa, DRC, with means to access free CD-ROMs and computers, to enable them to access health information. This indicates a lack of overall policy, strategy, and finance for ICT infusion within the country's higher-education sector (Lonji 2002).

Academic libraries have resorted to forming partnerships in order to establish formal methods for sharing information resources. In Southern Africa, the South African Bibliographic Network (SABINET) has been developed as a self-financing resource-sharing database. The database is based on a long-established process of contributions of cataloging records, provided by various libraries in the subregion, the majority of which are in South Africa, to form a union catalog. This collaborative effort laid a foundation for a culture of resource sharing and the establishment of a viable interlibrary lending service system. SABINET has evolved to offer fee-based online access, not only to its union catalog but also to a broker, for access to the OCLC databases and other online journal literature, for its members.

In Ghana, resource sharing efforts have been made since 1993 with help from donor funding: the Ghana Inter Library Lending and Document Delivery Network (GILLIDDNET). This is a partnership among five publicly funded academic libraries in Ghana and libraries in Denmark. The project is supported by Danida's Enhancement of Research Capacity program. (ENRECA) through the use of CD-ROM databases.

LIBRARY AUTOMATION

Sub-Saharan academic libraries embarked, gradually, on projects for the acquisition of technology to automate library operations to convert card catalogs to online public access. Assessment of the rate of library automation in select sub–Saharan libraries in the mid-nineties indicated that while academic libraries had plans for automation of library systems, full-scale automation of library operations was rare, on account of institution-wide shortage of funding.

Thus most of those academic libraries that have automated their library collections have been assisted, to a large extent, by donor funding for both hardware and software.

In East and South Africa, early beneficiaries of donor funding for the purchase of integrated library systems are Moi University in Kenya and the University of Zambia. The World Bank has been active in the financing of library automation through bilateral agreements and loans from the 1990s to date. The University of Dakar library in Senegal is one of the first sites that benefited, in 1996, through the provision of appropriate hardware and software under the World Bank project on Higher Education in African countries. The total budget for the project on "strengthening of Library services" was $17.4 million. It included the rehabilitation, extension, and equipping of the central library; acquisitions of books and periodicals, and the installation of an integrated library system. Recently the University of Conakry, in the Republic of Guinea, is reported to have signed a contract with the World Bank in 2002 for the strengthening of library services, including the installation of an integrated library system. Funding for the project is provided through a bilateral loan agreement between the World Bank and the national government (Sene 2004).

South African University Libraries have benefited from the post-apartheid reconstruction program, which is supported by various donors such as the joint initiative between the European Union and the Department of Education of South Africa and the Ford Foundation, Carnegie Cooperation, and Mellon Foundation. Hence there is a huge investment in integrated library automated systems, the benefits of which are maximized through consortia such as Gauteng and Environs Library Consortium (GAELIC). In addition the South African Site Licensing Initiative (SASLI) has been successfully established as part of a collaborative framework for integration of resource sharing and information services. Importantly, the large numbers of academic libraries and the government commitment to the revitalization of the formerly disadvantaged tertiary institutions in South Africa have made it possible to provide a country-wide capacity-building program for librarians.

The significance of these projects is that, through bilateral contracts, donors provide loans to governments for improvement of higher education. They demonstrate the governments' heightened awareness of the importance of academic libraries in the revitalization of higher education in African countries. It could be argued, of course, that the inclusion of library rehabilitation programs may be part of the donor's conditions (e.g., such as the World Bank's) rather than national interest. The case of Senegal, however, provides evidence that national interests determined priority areas that should be supported by such loan funding (Sene 2004). The availability of large sums earmarked for libraries is a welcome development that will have long-term impact on the ICT infrastructure in African universities, their academic libraries, and the information services they offer to support learning and research.

For most academic libraries in the late 1980s and 1990s, the strategy of investing in technology for the support of academic library information services was seen more as pilot services, facilitated by the use of free equipment and based on project proposals. It is hoped that the impact of these pilot projects will influence institutional decisions toward investment in more technological input. This assumption has been vindicated in the case of GILLDDINET, where university authorities not only commit to paying the monthly charges for the Internet service provider but also have begun to provide hardware to increase access to information through libraries. Sub-Saharan African countries, however, are many and have varied political interests and economic constraints that determine priorities. Therefore, what is applicable in one country is but an illustration that under certain conditions, investment by countries' universities into academic libraries is possible. The different contexts need to be researched in order to find out how university libraries can best advocate successfully for the provision of regular recurrent budgets to sustain donor-funded technological input, that empower librarians to provide adequate academic library support for research, teaching, and learning. Concern about sustainability has been commented upon as follows:

> When reviewing the innovations made in IT, it must be remembered that most, either wholly or in part derive from donor-provided support. . . . [W]hat happens when the aid stops remains a question mark.

ACCESS TO THE INTERNET

African academic libraries have lagged behind technological developments. Although Internet access has improved tremendously, as recently as 1994 Internet connectivity in Africa was said to be limited to four countries:

South Africa, Zambia, Egypt, and Algeria. As a result of low bandwidth, access to the World Wide Web is still very slow and therefore affects the quality of access to information. For example university libraries in Malawi reported that "in spite of the wide variety of electronic resources that are available through the PERI project and HINARI, the quality of the Internet connectivity is often a great hindrance" (Mwiyeriwa and Ngwira 2003, 9). Such experiences are not uncommon in the African contexts: They and have been referred to as the World Wide "Wait" experience of the Internet service.

Costs of access to information through the telecommunication infrastructures have added a burden to already overstretched university budgets. According to an ITU survey of 1999, the Internet monthly access charge as a percentage of the GDP per capita was, for example, in Uganda, $92 (107 percent) compared to the U.S. rate of $29 (1.2 percent). These factors initially influenced the rate of uptake of the Internet as an information access tool by African universities. It has been reported, however, that nineteen countries in Africa have since instituted low call charges for all Internet calls. Thus Internet access for delivery of information resources for research, teaching, and learning is improving as various interrelated policies are changed in favor of reducing cost of access and increasing both the bandwidth and access points.

The functionality of the Internet as a communication tool has encouraged the establishment of networks and partnerships within countries to widen access to various resources and to create local content databases on information scattered across various organizations.

CONSORTIA

As the twentieth century came to a close, the use of CD-ROM technology for delivery of information was the norm. CD-ROM is an appropriate and affordable technology for dissemination of individual titles of journal literature for developing countries. However, it is now being replaced by journal title aggregations, which consolidate journal subscriptions into a single database, accessible electronically at a consolidated subscription fee. The impact of this consolidation is that academic libraries with small budgets that can afford to subscribe to a number of relevant stand-alone CD-ROMs can no longer afford to continue subscriptions once the needed titles are aggregated in a stable of journals such as the EBSCO Host. Such service is of excellent value if aggregated journal titles match those required to meet information needs of an academic library information service. Generally, however, the available titles include many that are not specifically required for the support of either the curricula or research programs. It is this aspect that has encouraged the consideration of consortia agreements, which enable more than one library to share subscription costs and access. The process of negotiation empowers libraries to get lower subscription rates. Donors have once again provided assistance by facilitating the raising of awareness of strategies for negotiation with aggregators through the formation of partnerships and consortia.

An organization that has contributed to the formation of consortia is the Electronic Information for Libraries (eIFL), a project of the Open Society Initiative. It contributes by urging academic and national librarians to consider the potential for and benefits of consortia agreements with database aggregators for participating libraries within a country. It facilitates discussions across institutions on requirements and planning for successful partnership within a consortium and finances technological inputs when appropriate. Experiences from successful consortia such as SOLINET in the United States have been shared with emerging African consortia. There are now eleven countries in sub-Saharan Africa that are in the various stages of forming intrastate consortia: Angola, Botswana, Lesotho, Malawi, Mozambique, Namibia, Nigeria, South Africa, Swaziland, Zambia, and Zimbabwe. In Malawi the program financed the installation of VSATS to improve connectivity for university libraries in the Malawi Consortium (MALICO) (http:soros.epnet.com/eifl.description.asp [accessed May 5, 2004]).

The Programme for the Enhancement of Research Information (PERI) is another donor-funded program that seeks to alleviate problems that arise out of limited access to and lack of skilled use of the Internet. This program provides access to full-text electronic journals at preferential subscription rates for the benefit of developing countries.

PERI also strengthens publishing of African content and its accessibility to African libraries through the African Journals Online (AJOL) project. Abstracts are accessible free of charge to members of PERI, libraries in sub-Saharan Africa countries. The third components of PERI is capacity building for librarians in effective exploitation of the Internet as well as developing a pool of subregional in-service trainers, for sustainable local capacity-building skills.

Projects on consortium building facilitated by the eIFL and PERI have contributed to the development of various skills that are necessary for successful proactive approaches to the dynamic information environment, which must be managed for the benefit of individual users. The skills that have been learned and are in the process of being mastered are:

- Team building and coalitions
- Strategic planning for sustainable financing
- Development of cost benefit analysis approaches and skills operations

- Project management
- Advocacy skills.

Without ICTs and their influence on library services, consortium formation and networked services would not have been successful since, in manual operations, obstacles related to different space and time occupied by the various libraries would have been insurmountable.

The impact of the above-mentioned activities, however, has yet to be empirically assessed, while sustainability of the programs needs to be researched as a basis for strategic plans even as it is planned for within project proposals. Measuring the impact of these ICT-based programs on users is limited to quantitative data. Qualitative assessment requires well-thought-out research strategies that target knowledge output. It is clear, however, that if librarians are to assure sustainability of the spinoffs of ICTs in academic institutions, marketing of products and services as well as iterative evaluation of outcomes must be built into the pilot projects.

UNIVERSITY OF BOTSWANA LIBRARY: CASE STUDY

The University of Botswana was inaugurated in 1982 as the only university in a country with a population of less than two million. The growth rate of the student body has been rapid. In 1999/2000 there were eight thousand students, and by 2003/2004 there were more than thirteen thousand. In recognition of the relatively fast growth of the student population, the university committed to computerization of all facilities, including academic information services, as a strategic management tool. The university has established an information technology strategy that seeks "to enable the innovative and effective use of ICT to achieve the goals of the University" ("University of Botswana Information Technology Strategy"). These goals are underpinned by the vision that seeks to produce graduates who are independent, confident, self-directed critical thinkers in a technologically advanced teaching environment.

The University Library and the Computer and Information Technology Department are partners that are responsible for the information systems within the university. They share the provision infrastructure and support services for three clusters: teaching and learning, research and scholarship, and management and administration. The library is responsible for the support services of the first two clusters. This partnership has facilitated a holistic approach to policy making, budgetary provision and procurement strategies for both hardware and software, and sustainable technical support.

The efforts to automate library collections at the University of Botswana began in 1986. The installation of the TINLIB software of IME took place in 1989 (Mbaakanyi 1994). By 1996 the TINLIB software was fully operational. The following modules were in use: OPAC, acquisitions, serials, and circulation. In addition, twenty-five networked CD-ROM titles as well as limited e-mail facilities were on offer to users. In 1998 the university became concerned, as did the rest of the world, about the guarantees that all software in use would be year-2000 compliant. The difficulties in obtaining required assurances resulted in decisions for immediate migration to software that provided the Y2K assurance. The decision provided an opportunity for a deliberate search for a new integrated library system, based on terms of reference that took into account identified system needs and priorities that were agreed upon by the joint committee of the library and computer departments. The Innovative Integrated Library System was successfully installed by the end of 1999.

The unusual speed and capacity to implement this decision was influenced by the need to contain a situation that might have been disastrous for the core business of the university, if it had not been contained effectively. It demonstrated the effectiveness of the scare factor in successfully marshalling policy and financial support from university management to achieve the transition from one system to the other.

Assessment of ICT penetration in the university by 2003 indicates that 160 computers and sixteen OPAC terminals were accessible for use by students in the library. This is in addition to 869 computers available for student use in the various faculty laboratories (Ojedokun and Rao 2004). Library computer facilities were accessible for student use for 12.5 hours per day over seven days, while those in laboratories were accessible on average for seven hours per day over five days. The increased demand for twenty-four-hour access to library computer facilities is a challenge that needs to be explored in view of the shortage of skilled personnel to provide twenty-four-hour support and monitoring of computer usage.

The experience of the University of Botswana in the incorporation of ICTs to support academic library services has been similar to that of other university libraries in sub-Saharan Africa in most ways. The financing of collection development and hardware and software procurement, however, has been a significant exception. The library has over the last ten years been assured of annual recurrent budgets of between 5 and 10 percent of the total university budget and thus has sustained collection development. The infusion of ICTs continues to be budgeted for separately in support of the university's overall vision of the role of information and knowledge in higher education. Human resource development to encourage competency in the use of ITCs by all staff and strategic planning for emerging information services have been the greatest challenges of the time.

Information Literacy Skills Development

The University of Botswana has committed resources over the years to the development of communication and study skills for all first-year students. Librarians have been involved in teaching a tenth of the course, which was devoted to providing skills for effective use of library resources. The component focused on understanding computers and their use for accessing information resources, from grey print literature to OPAC and CD-ROM databases held by the library and those held elsewhere but accessible electronically.

In recognition of the need to encourage and develop resource-based learning systems, the university implemented compulsory general education courses in computer skills and information literacy for credit for all first-year students in the 2001–2002 academic year. The course is offered under the auspices of the Computer Science Department. Librarians are partners who facilitate learning and practical subject-linked workshops on all aspects of student information literacy skills development. The objectives of the computer and information skills (CIS) program are, among others:

- To promote and encourage computer and information literacy for all students
- To provide a basic-to-advanced computing and information competency that will allow students, regardless of their background, to be productive parts of the Information Society (Ojo 2003).

The Impact of ICTs

The impact of ICTs at the University of Botswana Library manifests in the change of focus from technical services to dynamic programs that facilitate:

- Increased electronic access to information resources: whether they are held locally through hard print CD-ROM or accessible through remote databases.
- Continuous capacity building and empowerment of library staff in effective exploitation of ICTs as a tool for support of learning and research programs.

While some library staff have embraced the dynamism and demands for new competencies brought about by ICT, others are intimidated by the requirement for them to develop competent practice. The fundamental impact of technology is that routine technical services in academic librarianship have been transferred to cadres that are efficient and effective in technical processing of library materials. Thus librarians' time is freed for the provision of analytical work that is based on regular interaction with faculty and students on curricula and the provision of subject/research information support services.

Library and information professionals today have a challenge and an opportunity in understanding and accepting the concept of proactive librarianship. Such an approach is essential for keeping up with the dynamic information environment. It further demands personal commitment to continuing education from librarians, if they are to meet the dynamic information needs of patrons effectively. This challenge may be daunting and stressful for some, while it empowers and fuels enthusiasm for others. The most notable impact of ICTs in academic libraries is thus the demand for a paradigm shift from static professional skills to dynamic reskilling, continuous up-skilling, and ongoing performance measurement in relation to service to users rather than a focus on the organization of information media. Training programs that address the management of change have not been developed systematically or financed adequately from within.

The University of Botswana has benefited from donor-funded capacity building projects for the mastery of the Internet and its effective exploitation for provision of information services. "The Internet Traveling Workshops" supported by the International Network for the Availability of Scientific Publications (INASP) have empowered local trainers. The spin-off is that skills filter down to librarians within the university and are shared with other librarians at the national level.

Another capacity-building project from which the University of Botswana has benefited uses a team approach to training. It is entitled "Use and Application of ICT Education in Africa" (UAICT-Africa). This is a joint venture by ten university libraries, one library science teaching department in Southern Africa and the Research and Development Department of Lund University Libraries in Sweden (Eriksson 2003). The main objective of the project is to set up an International Research Catalogue (IRC) on ICTs in education as well as to nurture a culture of collaboration through the Internet and demonstrate its capacity as a facilitator for access to global resource links and sharing of costs. The use of Listserv as a space for virtual meetings also contributes to the acceptance and practical use of the Internet platform for meetings in place of face-to-face meetings and conferences/workshops, which are expensive. The objectives of the project are yet to be fully attained, since the mode of learning is based on assumptions that participants are well motivated to commit personal time in fulfilling assignments; have good time-management skills; are good communicators and team learners; and, importantly, are capable of enjoying working independently. Details of the AICT-Africa project are accessible through netlab.lub.lu.se/sida/celi.

However, exposure to electronic means of communication is such a novelty to African communities that even in the banking industry—which is one of the front runners in the adoption of the information society "products" such as electronic transactions—research shows little awareness of essential good practices., For example in online banking:

> Internet service providers and banks have noted security as being very important, but online customers and clients are not as concerned . . . [because] educating customers about online security has been lacking and, despite South African banking being on par with international online business world, users are not fully clued up. (*Saturday Star* 2004)

This knowledge gap is highlighted as an example of the types of barriers that need to be consciously addressed by all concerned library professionals at policy and management levels, librarian operator levels, faculty levels, and user levels. If ICTs are to yield expected benefits and impact the role of administration in the management of change, librarians must institute mentoring and monitoring as good practice and continuous human resource development. The example above also demonstrates the difficulty of implementing change in management processes. Policy makers who are entrusted with customers' wealth seem to have failed to recognize the urgency for educating customers about their own role and responsibility in safeguarding their personal investment in being part of the information and knowledge society.

Development of ICT technical management/maintenance skills in academic libraries in sub-Saharan Africa has been recognized as being so important that training processes are well established. However, the scarcity factor continues to affect academic library IT projects negatively; skilled ICT personnel command high remuneration, and competition for their services is fierce. Still, they are indispensable for training users in the most efficient methods of exploiting information and knowledge resources in the IT industry.

The introduction of ICT in sub-Saharan Africa has been empowering. Library staff have strived to develop positive attitudes toward challenges posed by ICTs. Opportunities to explore ways of acquiring expertise in the exploitation of ICT have been offered to librarians through networking with donors, faculty, and IT professionals. The most significant impact of ICT has been the gradual acceptance of continuing education necessitated by the rapid changes that occur in the ICT arena and its application to information access and delivery. Thus library staff have gained skills for training users in the application of technology for the support of learning and research. They have gained users' support in advocating for improvement in the funding of library services. The impact of ICTs in academic libraries is well captured by Sene:

> The impact on the staff has been very important. All the staff, professional or not benefit from all possibilities given by the new environment, services and new tools. They have had opportunity to enhance their skills and knowledge. The change also gave them a new image of the profession and consequently has increased the value of the work made by librarians for the academic community and its role in the development and progress of the teaching and research. For the users this new library with all the new services available and possibilities offered by information technologies is now seen as an important pedagogic tool, like laboratories and other scientific equipments. (Sene 2004)

BIBLIOGRAPHY

Blair, R. D. D. 1991. *An Assessment of Progress and the Potential for Financial Diversification and Income Generation at selected African Universities. A Report to the World Bank.* Harare: Blair Management Services and SPECISS College.

Constructing Knowledge Societies: New Challenges for Tertiary Education. Washington, DC: The International Bank for Reconstruction and Development/The World Bank, 2002.

Coombe, T. 1991. *A Consultation on Higher Education in Africa. A Report to the Ford Foundation and the Rockefeller Foundation.* New York: Ford Foundation.

Eriksson, J. 2003. *INASP Newsletter* (November 2003): 13.

Kisenyi, M. 2003. "Collaborating to Preserve Kenya's Information Heritage." *INASP Newsletter* (November) 11: 6.

Lonji, C. 2002. "Access to Health Information For Medical Students in Kinshasa." *INASP Newsletter* (February) 2: 8–10.

Mbaakanyi, D. M. 1994. "Library Automation at the University of Botswana" in *Survival Strategies in African University Libraries: New Technologies in the Service of Information. Proceedings from a Workshop held in University of Zimbabwe,* edited by H. Patrickios and L. A. Levey. Washington, DC: AAAS.

Mutula, S. M. 2004. "IT Diffusion in Sub-Saharan Africa: Implications for Developing and Managing Digital Libraries." Keynote address delivered at the IFLA Africa section workshop on "Developing and Managing Digital Libraries" held in Gaborone, Botswana, February 23–27, 2004.

Mwiyeriwa, S. S., and M. E. Ngwira. 2003. "A Malawi Library Consortium: Learning as We Go." *INASP Newsletter* (November).

Ojedokun, A. A., and Kachiraju N. Rao, 2004. "The Learning Resource Center and Its Needs of Adequate Computing Facilities: The University of Botswana Library Experience" (unpublished article).

Ojo, S. O., et al. 2003. *Computing and Information Skills Fundamentals* 1. Gaborone: University of Botswana Department of Computer Science.

Patriokios, H., and L. A. Levey. 1994. *Survival Strategies in African University Libraries*. Washington, DC: AAAS.

Raseroka, H. K. 1992. "Microcomputer Applications for Managers," in *Proceedings of the First IFLA Africa Section Workshop Series, 7–16 December 1991*. Gaborone, Botswana: IFLA).

Sene, H. 2004. E-mail Communication to Raseroka, April 23, 2004.

World Bank. 1997. *Revitalizing Universities in Africa: Strategy and Guidelines*. Washington, DC: World Bank.

Chapter 9

The Impact of Technology on Library Collections and Services in Nigeria

C. C. Aguolu, I. Haruna, and I. E. Aguolu

ABSTRACT

This chapter is an overview of the application of modern technologies in the development and use of information resources and services in Nigerian libraries. It identifies the major technologies relevant to libraries and their use. The advantages of modern technologies to libraries are highlighted. Enhanced access to information resources, for instance, in the Internet and e-mail services, which have reduced the long-existing intellectual isolationism of Nigerian researchers, are discussed. Problems associated with full exploitation of computer and related technology services and their implications are discussed. The chapter concludes that library and information services in Nigeria are yet to transcend the traditional functions, as a number of library operations are still performed manually. As a way forward, recommendations are made on the basis of the findings.

INTRODUCTION

This chapter is not intended to be an empirical study as such but is essentially an overview of the state of the art in the application of modern technology to the development and use of information resources and services to satisfy the needs and demands of Nigerian scholars and researchers. Understandably, this is with special reference to university and other research libraries, where technology, albeit currently limited, has had an impact on both librarians and library patrons. However, an attempt is first made to identify the major technologies relevant to libraries and their use.

The earliest use of mechanical or semiautomated equipment in libraries the world over took place in the United States of America in the 1930s, with the introduction of punch-card procedures at the University of Texas. Other similar efforts in the use of punch cards by the Orders Unit of the Library of Congress followed in the 1940s. However, it was only in the 1960s when computers became widely available that use of modern technologies in libraries became widespread. In an age where labor costs have assumed astronomical heights and technological innovations occur ceaselessly and at reasonable cost, heavy dependence is placed on the application of technology to different spheres of human activities.

The inadequacies of the traditional library methods of handling scientific information in the United States were condemned in the 1940s by an American physicist, Vannevar Bush (1945), before the American Philosophical Society, where he contended that the problem of bibliographic and intellectual access to recorded information was "not so much we publish unduly in view of the extent and variety of the present day interests, but rather, that the publication has been extended far beyond our present ability to make real use of the record" (102). Almost sixty years after this caustic attack on the prevailing old-fashioned, ineffective information storage and retrieval devices in the United States, especially in science and technology, calling for the immediate application of modern technologies to information storage and retrieval problems, a similar situation now obtains in Nigeria, demanding a similar response through appropriate technologies, available and largely affordable.

Technological developments through the years have enabled people to perform a number of functions much faster and with less effort. Modern technologies in libraries create a new forum for global information access. Information technology encompasses all the electronic infrastructure and facilities, utilized by libraries to improve and provide quality services. Aina (2004) narrowly defines information technology (IT) thus:

> As an omnibus term that combines computers and telecommunications technology; hence it is sometimes called information and communications technology (ICT). It is concerned with the technology used in handling, acquiring, processing, storing and disseminating information. (301)

The use of computers and telecommunications technologies in information handling has arisen as a result of increased workload involved in coping with an information explosion with respect to Nigeria. Oyelude

(2004) has argued that the traditional library tools are limited in their ability to cope with the high degree of information retrieval and dissemination, and calls for the use of modern technologies.

A much wider definition of information technology than Aina's (2004) is that of the American Library Association (2003) which defines it as "the application of computers and other technologies to the acquisition, organization, storage, retrieval and dissemination of information" (12). By inference, therefore, computers provide the processing, storage, and retrieval, capabilities while telecommunications provide the capacity for the transfer and/or communication of data from one workstation to another, or from one individual to another.

In the development of library collections and the provision of effective services in the developing countries, these four categories of technologies could play and have historically played significant roles, namely:

- Microform, reprographic and printing/laser technologies;
- CD-ROM technology;
- Computer technology; and
- Telecommunications technology.

These are now available to librarians and largely affordable for use in promoting their information resource development and services to their clientele.

For example, micrographics (microfilm, microfiche, and microcard) present new ways of storing information more compactly and cheaply, especially information in books, newspapers, journals and technical reports. They have facilitated availability in Nigerian libraries of certain categories of information resources, no longer commercially available or out of print, to support teaching and research.

At the inception of many Nigerian university libraries, given a barely existing local publishing industry, micrographics were heavily used to build up local collections, especially the microfilm, microfiche, and microcard. The largest and first university library in Nigeria, at its commencement in 1948, quickly recognized the importance of developing its early book and serial collections with micrographics in a country then without any recognized academic publisher or bookseller. This was the University College Library, Ibadan, which, according to the first principal of the University College, on his purchase of three microfilm readers in 1948, possessed one microfilm reader more than the Bodleian University Library, Oxford. Again, on his acquisition of a reader printer in 1950, for retrieval of information from microcards, the university acquired the "first microcard reader in any library in the British commonwealth" (Mellanby 1958, 208–9).

Two major trends are now discernible in library computerization. The first is the employment of computers to perform repetitive and routine library tasks, while the second involves those applications meant to assist information retrieval. Aguolu and Aguolu (2002), as noted by Artandi (1972), Kimber (1974) and Rowley (1988), have observed that:

> Since the 1960s, computers have been used, to good effect, to perform many library functions from clerical housekeeping activities, such as ordering, cataloguing, serials control, and circulation of books and journals to reference and information services. (267)

The last service (reference and information services) relates to indexing and abstracting services, interlibrary loan services, current awareness services, networking for online interactive searching, and resource-sharing services.

One major aspect of computerization that has become handy for university libraries is the ability of computers to communicate with one another over short and long distances—computer networks. This brings in the added advantage of transfer and exchange of information. The position of computer applications to libraries in developing countries, as remarked by Akinyotu (1975), is, however, a far cry from that in developed countries. The reason he advanced is that it was not until the 1970s and 1980s that skeletal library applications of the computer became visible in developing countries, including Nigeria.

The present inquiry is to ascertain the extent to which the use of information technology (computers and their related technologies) assists in achieving higher efficiency in the provision of library collections and services to users, as well as the problems militating against their full implementation and utilization. University and other research libraries, especially of a special nature, were targets of early computerization in Nigeria. The *McGraw-Hill Encyclopedia of Science and Technology* (1997) stated that "a subject specialized research library acquires, classifies, catalogues, stores, and supplies both whole works and bits of information as precisely as possible, combining library practice with techniques of documentation or information science" (44–45).

IDENTIFIED TRENDS

The use of information and communication technologies (ICTs), particularly in library computerization, has been a topical issue in Nigeria for quite a considerable length of time. It has generated a number of seminars, workshops, and writings by writers such as Abolaji (2000), Adeyemi

(1991), Aguolu (1998), Aguolu and Aguolu (2002), Ajibero (1987), Akinyotu (1975), Alabi (1986), Dada (1999), Edoka (1969), Ehikhamenor (1990), Ifebuzor (1977), Lawani (1978), Mohammed (1999), NUC (1996), Nwajei (1987), Oketunji (2001), Olanlokun (1983), Olufeagba (1977), Ubogu (1989), and Ubogu and Gupta (1987). But not much actualization appears to have taken place. Library computerization in Nigeria seems to have remained at the pioneering stage for over two decades. To this extent, Oyinloye (2004), whose finding corroborates these writers, further asserted that, at the moment, most library operations such as acquisition, processing, information retrieval and reader services are still being performed manually in Nigerian libraries.

Computerization of library operations in Nigeria began in the 1970s at the University of Ibadan, Obafemi Awolowo University, and Ahmadu Bello University libraries. The first significant application of the computer to library functions in Nigeria was recorded in 1975, with the first issue of the University of Ibadan library's computer-produced Catalogue of Serials in the library. In 1988, Abubakar Tafawa Balewa University, Bauchi, computerized the acquisitions process of the library. In 1996, University of Ilorin library took remarkable strides in library automation leading to the computerization of some of its functions. This is equally true of the Ladoke Akintola University library, Ogbomosho. A number of the Nigerian university libraries, particularly the federally owned, have partially computerized their functions. This can conveniently be attributed to the fact that the National Universities Commission (1992), based on the advantages of TINLIB software over some like D/base IV, CDS/ISIS, X-Lib, Data Treck, Mastalib, Alice, and Glas, adopted same for twenty federal university libraries that benefited from the $20 million World Bank Loan Scheme for the rehabilitation of their fledgling structures and services. The TINLIB, which has long been discontinued and replaced by Alice for Windows, was adopted in order that Nigerian university libraries would have a standardized form of information exchange, using the same software and to equally facilitate their linkages with the Internet.

The preliminary survey conducted by these authors revealed that not one out of the twenty-three state-owned polytechnics has computerized its functions, though seven (38.9 percent) out of the eighteen federal polytechnics have very partially done so. The polytechnics are those of Auchi, Edo-State; Mubi, Adamawa State; Kaura-Namoda, Zamfara State; Ede, Oshun State; Kaduna, Kaduna State; Bida, Niger State; and Nasarawa, Nasawa State. It is, however, regrettable that none of the sixty-two Colleges of Education libraries (twenty federal, thirty-nine state, and three private) has demonstrated any genuine intention to computerize its library functions.

In 1984, the International Institute of Tropical Agriculture (IITA), Ibadan, changed from the traditional card catalogs and replaced them with a fully computerized integrated database system in 1989 with the use of Pattlelle Automated Search Information System (BASIS) Software. The IITA Library computerization involves the integration of acquisition, catalog, and circulation records in a database that can conveniently be accessed from seventy computer terminals located in and outside the library. In 1990, the Raw Materials Research and Development Council (RMRDC) library, having obtained version 2.3 of the mini-micro CD/ISIS from UNESCO, commenced the computerization process. The RMRDC developed the X-Lib software package for libraries. In 1991, full computerization of the Nigerian Institute of International Affairs (NIIA) library began when a systems analyst was employed and IBM computers with Tinlib software were acquired. These three (6.8 percent) research institute libraries, out of forty-four in Nigeria, are in the forefront of library computerization and in fact, in general application of information technology in Nigerian libraries and information centers.

A survey of academic and research libraries has revealed that only few of them use computers to automate technical services, to network operations like cataloging, authority control, compilation of bibliographies, and interlibrary loans. By inference, therefore, they use computers to perform repetitive, routine library activities such as production of catalog cards, ordering and accounting processes, processing serials, circulation control, and information retrieval. The CD-ROM technology, which has no doubt revolutionalized provision of information to users and publishing, is being used in the areas of cataloging, reference, and collection development units, but this technology is yet to be fully utilized in the libraries.

The Internet, on the other hand, is used by librarians for verification of new titles and placing orders from publishing houses. Electronic journals and newsletters are equally downloaded for the benefit of researchers. Computer networks have had tremendous positive impact on the level of customer satisfaction and patronage. One of the services enjoyed through computer network application is the Online Public Access Catalogue (OPAC), which enables users to access information. Automating the cataloging section has facilitated the OPAC, through which users can quickly and effectively search the computer-held files (database) of a library or group of libraries. The Internet, e-mail, and scanning facilities are offered in many academic libraries for fees. These generate income and equally help to ensure user satisfaction. A number of the libraries surveyed have computers. While very few have even partially computerized their services, others use computers mainly for word processing, in-house publishing, and administrative work.

DISCUSSION

Most academic and research libraries in Nigeria have not computerized any of their functions. The public card catalog and the visible index are still the finding tools for books and journals, respectively, in most libraries; likewise, indexes and abstracts are compiled manually. This finding coincides with that of Mohammed (1999), who commented thus:

> [The] majority of the libraries and information centres in the country are rarely ready to exploit the services of IT facilities such as the Internet, e-mail, telfax, computers, etc., for information acquisition, storage, retrieval and dissemination/transfer. (22)

Library and information services in Nigeria have yet to transcend the traditional functions.

The network services in Nigeria have led to huge reductions in the time spent by users to acquire needed information. This is crucial and a very good development, as time is one of the yardsticks with which services can be measured. In the advanced nations of the world, the computer is used in virtually all spheres of library operations. In Nigeria, however, though the need to apply computer techniques in library functions has long been acknowledged, the problems associated with exploitation of computer/IT services are many: finances, lack of expertise, efficient infrastructural facilities, spare parts, and computer knowledge or literacy of the users, and apathy on the part of the library authorities to computerization. Others include the erratic power supply of the National Electric Power Authority (NEPA) and unreliable communication links provided by the Nigerian Telecommunications Limited (NITEL). The libraries' dependency on NEPA and NITEL is to a great extent responsible for the inefficiency with which they provide services.

The nature of library personnel equally inhibits effective services and in turn slows users' pace of work. A good number of the personnel who have undergone the traditional form of training appear reluctant to embrace the current trends in improved library operations and services. This being so, Nigerian libraries have to adapt to modern trends in technology so that they do not lag behind the global village, occasioned by the imaginative use of modern information technologies.

As the information explosion continues to impinge on Nigerian society, as the body of human knowledge continues to shift direction, change relationships, and expand, and as the new technologies continue to evolve, the role of libraries in information provision will assume much greater significance. Therefore, in order to remain relevant in the new information age, Nigerian libraries and librarians have to explore new ways of rendering information services that are consonant with the demands and needs of the electronic age.

We are, of course, aware of the divergent views about the so-called information explosion resulting from these factors:

- The sheer volume of records of human communication generated daily;
- The diversity of the forms of these records (book, periodical, audiovisual, micrographic, machine readable);
- Their increasing complexity due to the ever- growing specialization and differentiation of knowledge; and
- The increasing use of many foreign languages in the publication of these records, especially in science and technology.

Globally, the library's traditional methods of information handling are unable to cope with the current information deluge, which cynical observers like Coblans (1971) and Wilson (1980) dismiss as "mere paper storm" or "paper inflation," not necessarily an explosion of information. Computer futurologists like Licklider (1965) and Lancaster (1978) contend that this information deluge can only be minimized by the immediate and drastic application of computer and telecommunication technologies. Coblans and Wilson maintain that much of what is published is repetitious or duplicative, based on a weak referencing system and editorial evaluation. Consequently, a vast number of journals add nothing to the growth of knowledge, being marginally relevant to researchers, duplicative in content, or not worth publishing in the first place. Colbans, then bibliometrically, posits that 90 percent of the world's significant scientific information can be found between 10 percent and 20 percent of the scientific literature.

Wilson (1980), who endorses Coblans's contention, submits that the growth of knowledge does not consist simply in

> the discovery of new items of knowledge but in the discovery of significant new items of knowledge. . . . [A] mere count of publications emanating from a research field tells us nothing except how busy the research field is. We must look at the content of the publications and decide whether that content represents an addition to knowledge. (4–5)

While Nigerian libraries have made meaningful use of microform, reprographic, and printing technologies in developing their collections, especially in the early years of evolution, the potentials of CD-ROM (Compact Disc–Read Only Memory) technology—a new technology not requiring efficient telecommunication, that makes available relevant information for teaching and research—are yet to be adequately exploited. The particular suitability of CD-ROMs for information requiring large storage capacity, color graphics, and sound and video components, has revolutionalized publishing. The CD-ROM is an optical disc with a prodigious capacity of about 700 floppy discs—enough memory to store about 300,000 text pages. Data from CD-ROMs are read by the laser beam used for recording them. While the user can read the disc, he cannot write on it or erase anything written on it.

Being computer readable in form, CD-ROM can be read using a computer and the appropriate software. Many publishers in Europe and the United States from which the Nigerian academics and researchers purchase the bulk of their books and journals now publish simultaneously in both print and CD-ROM formats. Publishers of reference books, particularly encyclopedias, dictionaries, travel guidebooks, directories, and journals, are shifting to the CD-ROM format. A few Nigerian university and research libraries subscribe to CD-ROM but primarily as stand-alone informational units to be counted like books and journals, and do not yet network electronically with other libraries.

Nevertheless, through the Internet and CD-ROM technologies, the Nigerian researchers have remarkably enhanced their literature awareness in research and publication, for lack of which scientists in the developing countries have frequently been faulted by the international journals (Freeman 1969; Gordon 1979). A well-designed computer-based national information network, with its hub either at the National Library of Nigeria at Abuja or at the National Universities Commission in the same city, will link up all Nigerian major libraries—national, special, public, and academic or research—with the world's major libraries. This will ensure availability of adequate information resources in various media and format for teaching and research and their effective use by the clientele of the cooperating libraries, regardless of their locations, to satisfy user needs and demands.

PROSPECTS

The computer is an essential tool for librarians to improve their services to their clientele. They must reject, as Alemna (1996) argues, "the notion that automation will take over their functions and hence their usefulness" (68). Alemna further contends that this notion may result in a failure to rise to the challenges posed by computerization and the application of other technologies to libraries. Aguolu and Aguolu (2002) have thus noted:

> By computerizing clerical and repetitive functions in acquisition, serials control, circulation and cataloguing, and by fitting appropriate telecommunications

gadgets to computers, thus facilitating local, regional, national and worldwide accessing of numerous bibliographic, numeric, textual databases for scholars and professionals, librarians will be enhancing their job performance in information handing. (273)

Computerization has facilitated the creation of local files and databases—thus making it easier and quicker to search and retrieve information. With the advent of microcomputers, a number of libraries have demonstrated their genuine intentions to mount small databases in subject fields on computers.

At present, acquisition of electronic information resources to complement and supplement printed and other traditional information resources in various subject fields, especially in science and technology, has been limited and largely idiosyncratic. For example, the CD-ROM technology, which is quite amenable to a local, regional, and national computer-based information network configuration for developing and sharing information resources, is yet to be tapped by the Nigerian libraries to enhance access to information needed and demanded by students and lecturers.

It was against this background that the National Virtual (Digital) Library concept was nurtured and is being implemented by the Nigeria's Federal Ministry of Education. This Virtual Library, also known as an electronic or digital library, will utilize computer and telecommunications technologies to make access possible to various information resources generated locally and abroad. The National Virtual Library Project, which will involve all the universities, colleges of education, polytechnics, lower levels of the educational system, and public libraries, has begun to be implemented with eleven universities. The implementation of this project is in five phases. A model virtual library will be set up at the National Universities Commission (NUC) at Abuja, to serve as the hub of the university-based virtual libraries.

The major objectives of the first phase of this National Virtual Library project are as follows:

- To improve the quality of teaching and research in institutions of higher learning in Nigeria through the provision of current books, journals, and other library resources;
- To enhance access of academic libraries serving the education community in Nigeria to library and information resources;
- To enhance scholarship, research, and life-long learning through the establishment of permanent access to shared global archival collections;
- To provide guidance for academic libraries in applying appropriate technologies used in the production of digital library resources; and

- To enhance the use and usability of globally distributed networks of library resources. (Nigeria Federal Ministry of Education n.d.)

The delivery service of the National Virtual (Digital) Library will be through the Internet, CD-ROMs, and Wide-Area Network (WAN), and the information resources to be made available for accessing and downloading for teaching and research will comprise mostly current books, journals, theses/dissertations, databases in different subject fields, maps, photos, and videos. The mission of this virtual library project, succinctly put, is to provide enhanced access to national and international resources and to share locally generated and available information resources with the international community.

Since the commencement of this project about two years ago, some university and other research libraries have been galvanized into action, setting up Internet services, computerizing some aspects of their library operations and services, retraining their library staff to acquire the necessary skills or competence for computer-based services, and acquiring electronic information resources in addition to the traditional information resources, to ensure the availability of needed materials for their potential researchers.

More importantly, scholarly communication between scholars or researchers in Nigeria and the advanced countries such as the United States and Great Britain has significantly improved. University students and academic staff now enjoy enhanced access to information resources in the Internet, either downloaded to be read on the computer screen or printed out as hard copies for use at convenience, and e-mail services to interested scholars abroad working in similar or identical research fields, sometimes leading to collaborative work and access to current journal citations or full-text papers. This, in effect, has reduced the long-existing intellectual isolation of Nigerian researchers, especially in science and technology.

The Internet is becoming increasingly universal for effective distribution of book titles in the world markets. Amazon Co. and Bertelsmann AG of Germany have become the world's largest booksellers without having physical bookshops, as orders can be made over the Internet under specified conditions. Nigerian librarians can equally harness, to the improvement of their library and information services, the modern developments in computer, telecommunication, reprographic, printing and micrographic technologies. By implications, therefore, the future of library and information services in Nigeria and worldwide is bound up closely with the development of information technology, as many operations can be enhanced and some new ones developed, using appropriate technologies.

CONCLUSION AND RECOMMENDATIONS

Information technology—comprising e-mail, computer faxing, teleconferencing, video-conferencing, provision of up-to-date information, and improved professional and corporate image—has a pervasive impact on library operations in terms of accuracy, elimination of drudgery, speed, and enhanced communication. Ceaseless advances in modern technologies the world over equally mean greater realization of the goals of our libraries as they have drastically raised output and lowered costs.

Information technologies are a well-established feature of modern libraries in developed countries, but the financial, infrastructural, technical, and staffing inadequacies that characterize developing countries impede widespread use of computers and their related technologies in Nigerian libraries. These have kept Nigeria at the rudimentary stage of the electronic library.

Since libraries' funds do not appear adequate for executing the range of services expected of them, it is recommended that governments and parent bodies of the surveyed libraries increase the allocation given to them (the libraries) so that services can be provided to meet manifest needs. Nigerian libraries and librarians have to explore new avenues through which information services rendered are consonant with the demands and needs of the twenty-first century.

BIBLIOGRAPHY

Abolaji, J. A. 2000. "Automation of Cataloguing Process in Nigerian Libraries: The Experience of Hezekiah Oluwasanmi Library, Obafemi Awolowo University, The-Ife." *Nigerian Libraries* 34, no. 2: 1–7.

Adeyemi, N. M. 1991. "Issues in the Provision of Information to Developing Countries." *African Journal of Library, Archival and Information Science* 1, no. 1: 1–17.

Aguolu, C. C., and I. E. Aguolu. 2002. *Libraries and Information Management in Nigerian: Seminal Essays on Themes and Problems.* Maiduguri, Nigeria: ED-Linform Services.

Aguolu, I. E. 1998. "Library Computerization in Nigeria: Towards a Steadier Progress." *Lagos Librarian* 25, nos. 1 and 2: 8–13.

Aina, L. O. 2004. *Library and Information Science Text for Africa.* Ibadan, Nigeria: Third World Information Services.

Ajibero, M. I. 1987. "Factors Affecting the Attitude of Librarians toward Media Technologies in Nigeria Universities." *Lagos Librarian* 14, no. 1: 65–71.

Akinyotu, A. 1975. "Computerization of Library Processes in Nigeria." *Nigerian Libraries* 12, nos. 1–3.

"American Library Association." 2003. In *Encyclopedia of Library and Information Science,* vol. 20, edited by Allen Kent et al. New York: Marcel Decker.

Alabi, G. A. 1986. "Library Automation in Nigeria Universities." *Information and Development* 2, no. 3: 42–49.

Alemna, S. 1996. *Issues in the African Librarianship.* Accra: Type Co.

Artandi, A. A. 1972. *An Introduction to Computers in Information Science.* New York: Scarecrow Press.

Bush, V. 1945. "As We MayThink." *Atlantic* Monthly, no. 176: 102.

Coblans, H. H. 1971. "An Inflation of Paper: A Note on the Growth of Literature." *Herald of Library Science* 9: 349–56.

Dada, G. 1999. "Computer Networking in National Libraries." *Lagos Librarian* 20, nos. 1 and 2: 1–5.

Edoka, B. E. 1969. "Prospects of Computer-Based Circulation System in Nigerian University Libraries." *Nigerian Library and Information Science Review* 1: 115.

Ehikhamenor, F. A. 1990. "Automation in Nigerian University Libraries: Progress or Mirage?" *Nigerian Library and Information Science Review* 8: 1.

Freeman, C. 1969. *Measurement of Output of Research and Experimental Development: A Review Paper.* Paris: UNESCO.

Gordon, M. 1979. "Deficiencies of Scientific Information Access and Output in Less Developed Countries." *Journal of the American Society for Information Science* 30: 340–42.

Ifebuzor, C. C., et al. 1977 "Automation and Library Cooperation for Nigerian Libraries." *Nigerian Libraries* 1: 1–3.

Kimber, R. T. 1974. *Automation and Libraries.* Oxford: Pergamon Press.

Lancaster, F. W. 1978. *Toward Paperless Information Systems.* New York: Academic Press.

Lawani, S. M. 1978. "Modern Technologies and Library Services." *Nigerian Libraries* 2: 1–3.

Licklider, J. C. R. 1965. *Libraries of the Future.* Cambridge, MA: MIT Press.

McGraw-Hill Encyclopedia of Science and Technology, 2nd ed. 1997. New York: McGraw-Hill.

Mellanby, K. 1958. *The Birth of Nigeria's University.* Ibadan, Nigeria: Ibadan University Press.

Mohammed, Z. 1999. "Automation and Internet in Nigeria Library and Information Centres: Obstacles, Prospects and Strategies." *Lagos Librarian* 20, nos. 1 and 2: 20–25.

National Universities Commission. 1992. "TINLIB in the Services of Nigerian University Libraries." *World Bank Project News* 1, no. 3: 4–5.

———. 1996. "Nigerian University Library System." *Library Bulletin* 1, no. 1: 4–7.

Nigeria, Federal Ministry of Education. n.d. *National Virtual Library of Nigeria.* Abuja, Nigeria: Ministry of Education.

Nwajei, E. F. 1987. "The Use of Computers in Nigeria Libraries: Experience and Prospects for the Future." *Nigerbiblios* 12, no. 2: 8–14.

Oketunji, I. 2001. "Computer Applications to Libraries." In *Libraries and Librarians: Making a Difference in the Knowledge Age: A Compendium of Papers Presented at the 39th Annual Conference and AGM.* Sam Mbakwe Hall, Concord Hotal, Owerri, June 17–22, 2001.

Olanlokun, S. O. 1983. "Micrographics in Libraries: Utilization and Trends." *Lagos Librarian* 10, no. 1: 18–27.

Olufeagba, B. J. 1977. "Computer and Circulation Control." *Nigerian Libraries* 13: 1–3.

Oyelude, A. O. 2004. "Academic Libraries: The State of the Art." In *Technology for Information Management and Service: Modern Libraries and Information Centres in Developing Countries,* edited by E. C. Madu. Ibadan, Nigeria: Evi-Coleman Publications.

Oyinloye, A. M. 2004. "Software Packages for Libraries in Nigeria." In *Technology for Information Management and Service: Modern Libraries and Information Centres in Developing Countries,* edited by E. C. Madu. Ibadan, Nigeria. Evi-Coleman Publications.

Rowley, J. E. 1988. *Computers and Libraries.* London: Clive Bingley.

Ubogu, F. N. 1989. "Development of a Computer-Based Library and Information System: ISIS Opinion." *Nigerian Library and Information Science Review* 7, no. 2: 24–31.

Ubogu, F. N., and D. K. Gupta. 1987. "Mini and Micro-Computers: Underused and Possible Use in Nigeria Libraries, Documentation and Information Centres." *Nigerbiblios* 12, no. 2: 17–23.

Wilson, P. 1980. "Limits to the Growth of Knowledge: The Case of the Sound and Management Science." *Library Quarterly* 50, no. 1: 4–5.

Chapter 10

Bridging the Technological, Language, and Cultural Gap: Partnering with an Academic Library in Francophone Africa

James J. Natsis

ABSTRACT

West Virginia State University received funding from the International Development Partnerships grant under the agreement between the United Negro College Fund and the U.S. Agency for International Development to engage a three-year project with the Université Nationale du Bénin in Benin, West Africa. Using the two academic libraries as focal points, the project increased human and informational capacity at both universities, expanded information services, forged leadership in the communities, and formed a substantial and sustainable bridge between the two societies.

INTRODUCTION

The purpose of this chapter is to report on and analyze a three-year project between the West Virginia State University (WVSU) Drain-Jordan Library and the main library of the Université Nationale du Bénin (NUB) in Benin, West Africa. The project was funded by an International Development Partnerships grant under the agreement between the United Negro College Fund and the U.S. Agency for International Development (USAID).

The three-year project increased human and informational capacity at both WVSU and the NUB and technological capacity at the NUB. Using

the two academic libraries as focal points, the project expanded information services, forged leadership in the communities, and formed a substantial and sustainable bridge between the two societies. The project also
addressed the objective of "women in development" and increased foreign language capacity at both institutions.

BACKGROUND

West Virginia State University

West Virginia State University, located in Charleston, the state's center of
government, industry, business, and population, is the largest institution of
higher education in the Kanawha Valley and serves as a major resource center for this metropolitan area. WVSU, through the administratively linked
West Virginia State Community and Technical College (WVSCTC), offers
training and retraining for workforce development, basic and literacy education, occupational and associate degree programs, developmental and
continuing education, and transfer preparation. The undergraduate schools
of WVSU offer comprehensive and distinguished baccalaureate degree programs in business, liberal arts, professional studies, sciences, and social sciences. In addition, WVSU recently implemented its first graduate programs
and underwent status change from a college to a university.

Founded in 1891 as a land-grant institute, West Virginia State University attained national prominence as an institution of higher education for
blacks and continues to serve as a center of black culture. Voluntary de-
segregation in 1954 created a distinctive character that inspired the motto
"A Living Laboratory of Human Relations." The university has the most
racially and culturally diverse student body, faculty, and staff of all institutions of higher education in the state of West Virginia.

International Studies

West Virginia State University has been welcoming international students, teaching international courses, and hiring international faculty for
almost a century. These efforts, however, have intensified in recent years
in acknowledgment of the university's increasingly important role in the
social and economic development of the Greater Kanawha Valley. The
university has always been committed to cooperative planning with business, labor, government, and educational and social organizations in the
region. A recent emphasis is on recognizing the state's escalating development and planned linkages with other nations. For example, since 1992,
twenty-five international faculty/staff have been added, modern foreign

language offerings have expanded, and students from Brazil, India, Japan, Thailand, Spain, South Africa, Sierra Leone, Ghana, Germany, Australia, and other countries have attended the University. These numbers, although modest, indicate important steps toward internationalizing campus life for a student body of five thousand commuters and on-campus residents. In concert with these efforts, an informal Faculty International Advisory Committee (FIAC) was organized in 1997 to begin the work of coordinating an international curriculum.

Title VI Grant

One of the major accomplishments in internationalizing the university was the U.S. Department of Education (DOE) award in 1998 under Title VI: Undergraduate International Studies and Foreign Language Program. The grant provided funds for a project entitled "Strengthening International Studies at an Historically Black University in Appalachia." As a result of the grant the International Studies Program at West Virginia State University was established with an office and full-time coordinator. The focus of the grant was the region of French West Africa and, more specifically, the Republic of Benin. Working in partnership with the Université Nationale du Bénin (National University of Benin, or NUB), selected faculty studied the language, culture, and economy of Benin and developed new courses and units in previous courses reflecting a global perspective. Faculty participants developed a new minor in International Studies that is now promoted through a newsletter and campus events. More important, thanks to the partnership initiated with the NUB as part of the Title VI grant, the university received an International Development Partnerships (IDP) grant in the amount of $250,000[1] from the United Negro College Fund and the U.S. Agency for International Development to engage in a three-year project (1999–2002) linking the libraries of the two institutions. The IDP grant focused on an exchange of librarians on a rotating basis, the acquisition of books and technological resource materials for the NUB library, and the opening of the Benin Room, a special collection of Benin government documents, literature, and artifacts, at the WVSU Drain-Jordan Library.

West Virginia State University Drain-Jordan Library

The Drain-Jordan Library has been an important part of West Virginia State University since 1951. During the past fifty-three years, the library has supported the curriculum and research needs of faculty and students by developing collections and introducing changes, including the technology to keep up with the developments in the field. One of the major changes since 1996 has been emphasis on technology and access to resources such as journals through online databases. Although the library subscribed to only 495 journals in 2003, students, faculty, and all other users had access to over 22,647 journal titles through various Web-linked databases.

According to a survey conducted by the Office of Student Affairs, 71.2 percent of students used the library facilities during the academic year 2002–2003. The library tops the ten most frequently utilized services on campus. In addition, 87,917 searches were conducted by users on various databases in the library during the year as compared to 20,966 during 2001–2002.

THE COUNTRY OF BENIN

Benin is located in West Africa and is bordered on the north by Burkina Faso and the Republic of Niger, on the east by the Federal Republic of Nigeria, and on the west by the Republic of Togo. A relatively small country in African terms, Benin covers an area of 43,483 square miles, or 112,600 square kilometers, slightly less than the state of Pennsylvania. Although the capital is Porto Novo, the largest city and port is Cotonou. Both cities are located in the southern coastal region.

Many languages are spoken in Benin, but French remains the official language, even though full independence from France was achieved in 1960. The economy of Benin, with a population of more than six million habitants, remains underdeveloped and dependent on subsistence agriculture, cotton production, and regional trade. Literacy rates are low—the average literacy rate is 37 percent, while only 25.8 percent of women are literate. These conditions of limited economic development and low literacy rates led to the idea for the proposed project described below.

Université Nationale du Bénin (National University of Benin)

The National University of Benin[2] was created in 1970. It unites the body of higher education establishments of Benin spread over different campuses in six different cities: Abomey-Calvi (seat of the executive office, seat of the university library management, and six establishments), Cotonou (five establishments), Porto-Novo (two establishments), Lokossa (one establishment), Parakou (one establishment), and Natitingou (one establishment).

The National University of Benin Library

The NUB University Library is situated on the Abomey-Calavi campus. It offers varied services to students and professors in support of their educational, instructional, and research activities. The library holds approximately fifty thousand books and a small number of mainly noncurrent periodicals.

Because of its importance at the heart of Benin society and to the public that it serves, the NUB Library is frequently solicited to assist schools and various associations outside of the university. Through its assistance to these outside educational institutions and associations, the NUB Library contributes to the government's carrying out of its educational objectives for the population. This is a difficult mission, because, despite its contributions, the library is plagued by certain difficulties persisting in, among other areas, the following:

- The state of documentary resources
- The state of its infrastructures
- The state of its human resources
- The state of technological resources.

The library's collections are insufficient in quantity and lack variety, in particular, for documents on the *research* level. Periodicals give only the basic essentials, while updated documents are almost nonexistent.[3] The premises are ill adapted to the volume of the clientele and to the nature of the offered services. The documents, much like the personnel, are poorly lodged.

Several years ago, the NUB Library began the process of collecting information through web links and databanks. However, the rhythm and the efficiency of these efforts were limited partially by financial constraints as well as by a lack of qualified personnel. The project with WVSU assisted the NUB Library in redressing some of these deficiencies and thus increasing its effectiveness in serving the Benin people's needs.

It is also worth noting that, in addition to the NUB Library, there are several other specialty libraries within the NUB system—e.g., business, agronomy, medical—located on the Calavi campus and elsewhere. The Campus Numérique de la Francophone de Cotonou (CNFC) occupies a wing of the NUB Library but is funded and managed separately.[4] The CNFC was expanded during our project period to include more than five dozen computers and a number of computer laboratories equipped for training and workshops. The Belgium government also added ten computers to the NUB Library during our project period. The specialty libraries, the CNFC, and the Belgium government contribution fall mainly outside of the scope of the WVSU-NUB IDP Partnership.[5]

LANGUAGE AND CULTURAL CHALLENGES

West Virginia State University sent six participants to Benin as part of an exchange. The Coordinator of International Affairs[6] and the Drain-Jordan

Library Director made a number of short-term planning trips, while three librarians spent one to two months each in Benin. The WVSU Computer Services Director also visited the NUB campus to consult with the project directors regarding the acquisition and installation of computers and network services. The NUB sent a total of six participants to West Virginia as well. The NUB Library Director made several short-term planning trips, while five librarians spent two months each in West Virginia. Among the eight librarian exchangees, five were women: two from WVSU and three from the NUB.

The state of West Virginia has one of the highest percentages of English-only speakers in the United States. The state also has a very small immigration population. Librarians at the Drain-Jordan library (with one exception) had virtually no experience in international travel and no background in the French language. In fact, there was an overall resistance to embracing another culture, which created a difficult environment for recruiting participants to take part in the exchange and for receiving Benin counterparts and speaking to them in their more familiar French language. Although most educated Beninois (or Beninese) have studied English to some extent in school, most have very little need for English in their day-to-day activities. They have had even less exposure to American slang and an American accent.

On the cultural front, Benin participants found the pace of life and private space in the United States to be somewhat of a challenge. They also had difficulties adjusting to an American diet. WVSU participants found the concept of time to be interpreted differently in Benin.

A former WVSU acquisitions librarian who spent two months in Benin as part of the exchange illuminates some of the nuances involving language and cultural differences. He reports, "Because many of the publications of West Africa are written in French, the Africana librarian would be wise to learn the French language, at least to a basic level." He goes on to say that

> [o]ur foundational belief systems, ways of life, methods of accomplishing tasks. . . . [A]ll of these things are fundamentally different. So, if you plan to start a thorough collection of Africana materials in your library, you will need to be far more flexible and non-traditional in your thinking when dealing with foreign publishers and dealers.

PROJECT OBJECTIVES AND ACTIVITIES

The three-year project increased human and informational capacity at both WVSU and the NUB and technological capacity at the NUB. Using the two academic libraries as focal points, the project expanded information services, forged leadership in the communities, and formed a substantial and sustainable bridge between the two societies. The project also addressed the objective of "women in development" and increased foreign language capacity at both institutions.

Women in Development

Among the eight librarians participating in the exchange, five were women. This was an important objective, and thus we had to aggressively insist that the NUB find qualified women librarians available and willing to travel to West Virginia for two months. One of the difficulties in finding qualified women was that, contrary to the nature of the library profession in the United States, where women librarians dominate, although women are well represented in the libraries of Benin, most work as staff because very few have been educated as professional librarians. The lack of professional women librarians in countries such as Benin makes this type of project, one that involves women, difficult, yet that much more important.

We held a symposium in April 2001 at WVSU titled "Literacy, Women, and Libraries." The NUB Library Director was present to educate participants, mainly from West Virginia libraries, as to the challenges women face in the library profession in Benin. The luncheon keynote was delivered by the coordinator of the African Section at the Library of Congress.

Leadership Development

The project enabled participants to develop library skills (in both public and technical services) and empowered them to train, motivate, guide, inspire, and direct others. Additionally, both institutions served as leaders in their respective communities through workshops and seminars. This was especially the case at the NUB, where librarians from various establishments attended a seminar designed to discuss and analyze librarianship in Benin.

Publicity was wide ranging (see below), and project directors met with library, administrative, and political leaders in Benin to promote the project. This greatly improved the image and exposure of the NUB Library. And finally, development of the overall technological, human, and material capacity at the NUB Library as a result of the three-year project enabled it to better carry out its role as a leader in librarianship in the country.

Bridging Programs and Relations

The project connected the WVSU and NUB libraries through knowledge exchange of information systems and the creation of a public archives for Benin (the Benin Room) at WVSU. This achievement indirectly assisted the USAID's Mission in Benin to increase literacy among its population.

Benin nationals on exchange at WVSU visited classes and local schools and held lectures. U.S. visitors at the NUB exposed many people to American culture. Friendships have been forged that will last for many years. WVSU also established a rapport with the Benin embassy in Washington, DC, as the Benin ambassador visited the WVSU campus on two occasions to promote the partnership.

Public Relations

The project has been acknowledged by the Minister of Higher Education in Benin, the Benin ambassador, and other high officials in Benin. WVSU administrators were also very supportive of the partnership. This was evident at the First International Banquet (1999) and at the Benin Room opening ceremony (2002). The latter event was attended by the Benin ambassador, the U.S. Department of Education Title VI program director, the UNCFSP IDP program director, the WVSU president, WVSU vice presidents, and many others.

The project attracted much media attention both in West Virginia and in Benin, and a number of scholarly articles and presentations were produced. Newspaper and weekly publication articles appeared in the *Charleston Gazette*, the *Charleston Daily Mail*, the *Beacon Digest* (WV), *7 Jours*, *La Nation*, *La Républicain*, and *Matinale*.

The following journal articles were produced from the project:

R. N. Sharma and Jeannie Bess. August 2000. "West Virginia to West Africa and Back: An International Collaboration." *American Libraries*

R. N. Sharma. April/June 2001. "Literacy, Women and Libraries: An Intercontinental Collaboration." *West Virginia Libraries*

R. N. Sharma. July 2001. "Literacy, Women and Libraries: An Intercontinental Collaboration." *Library Times International*

R. N. Sharma. July 2001. "Benin: NUB Hosts International Seminar on Documentation Cooperation." *Library Times International*

John M. Kistler. 2003. "Special Acquisitions: Collecting African Materials." In *Acquisitions in Different and Special Subject Areas*, edited by Abdulfazal Fazal Kabir, and published by the Haworth Press.

Increasing Capacity

It was a challenge for a small university library with limited resources like the Drain-Jordan Library to aid a library in the developing world.

Overcoming this challenge instilled a new confidence in the library leadership. The development of the Benin Room was a slow process but was nonetheless completed on time. It adds a new dimension to the library, as the Benin Room has expanded beyond the four-wall enclosure to include permanent displays on Africa that now occupy the northwest corner of the first floor.

The project reinforced international programs at WVSU and has added to the international mission at WVSU of "increasing global awareness." The addition of new books, computers, photocopy machines, and improved Internet connections has greatly increased capacity at the NUB.

Human Capacity

Librarians at both WVSU and the NUB have increased their professional skills. Through the partnership with WVSU, NUB librarians and the NUB Library director have been brought up to date regarding technology and information literacy. They are much better equipped to face challenges in an ever-changing world.

Approximately fifty librarians and university administrators participated in the two-day symposium "Literacy, Women, and Librarians: An International Collaboration." The symposium was held in April 2001 on the WVSU campus. A seminar was held on the NUB campus in June 2001 entitled "Documentation et Cooperation Documentaire à l'Université Nationale du Bénin." Thirty-six librarians from various libraries met to discuss the state of librarianship in Benin. A six-day workshop was held toward the end of the project at the NUB Library to train librarians and professors to use computers and network equipment.

Informational Capacity

WVSU maintains a special collection of Benin archives and publications. This collection is updated yearly, mainly through the library's general acquisitions. The NUB Library intended to serve as a liaison in the procurement of publications directly from Benin. Unfortunately, it is often difficult to move and motivate bureaucrats who are in charge of keeping and distributing these documents and publications.

The NUB Library collection was updated during the project with several purchases of new books. These books were selected by NUB Library staff according to the needs of their library. The project also developed an online central library catalog for the NUB Library and a bilingual website: www.collegetec.com/benin.

Technological Capacity

The NUB Library computer service director spent quality time with computer services staff at WVSU and visited several libraries in the United States to improve his technological capacity. NUB Library staff also received training in the use of computers and informational network systems. The project website, computer network development, photocopy machines, fax machine, and new computers all contributed to boosting the NUB Library technological capacity.

STRENGTHENING LANGUAGE DEVELOPMENT

The project budgeted sufficient funds for tutoring librarians months in advance of their overseas travel and during their stay within the country according to individual needs. Nonetheless, language presented a challenge, as it was difficult to train WVSU participants to learn French—a language that they had very little exposure to—before departing for Benin, and none of the librarians spoke French to aid arriving Benin librarians. In addition, although several of the Benin librarians spoke and understood English fairly well before arriving, others had difficulties. Fortunately, the project director at WVSU spoke fluent French, which facilitated contact with arriving Beninois librarians, and allowed for smooth communications during planning trips to Benin.

LESSONS LEARNED

Overall consensus of the partnership was quite positive. We were able to capture the attention of authorities at WVSU and, perhaps more important, of NUB leadership and national officials in Benin. This was important, as the NUB Library director seeks support and funding to sustain and improve the institutional capacity at the NUB Library.

The transfer of money presented many challenges, and we had to learn along the way how to best deal with this matter. Checks took many months to clear, so we concluded that money wire transfers would be more practical. Once the route for these wire transfers was established (via Europe), we were able to resolve the problem for the most part.

We also felt it was important to purchase computer equipment that was suited for use in Benin: French-language software and keyboards configured for French language use. We felt that purchasing computers in country ensured that the equipment was properly suited to that country's needs and that repairs and maintenance guarantees would be respected.

We also ordered books directly from a local distributor. In-country purchases offered an additional advantage, as it provided commerce to local vendors.

And finally, cross-cultural and language differences will always be an issue in this type of project, especially in cultivating mutual respect, whereby one language or culture does not necessarily dominate.

CONCLUSION

In retrospect, the challenges that such a project presented were constant and at times quite daunting. Nonetheless, the project team adapted to the difficulties it faced and eventually met its objectives. Working in the developing world under any circumstances is tiring and frustrating. French-speaking Africa offers increased challenges because: (1) it is Africa; (2) the francophone countries share systems (education, documentation, electricity, communications, etc.) that are not compatible with those found in the United States; and (3) the French language is deeply infused throughout these systems and thus English is not readily accepted.

The people are nonetheless friendly and hospitable. The local African traditions and cultures, juxtaposed with a francophone/francophile perspective on the world, offer a fruitful and insightful experience.

NOTES

1. The original International Development Partnerships (IDP) grant for $200,000 was supplemented with a special Education for Democracy and Development Initiatives (EDDI) grant for $50,000.

2. Precise language is important with regard to the university's name. The English-language "University of Benin" is located in Benin City, Nigeria. The French-language "Université du Bénin" is located in Lomé, Togo. The French-language "Université *Nationale* du Bénin"—our project partner—is located in the Republic of Benin. For the purpose of the project and consequently for this chapter, we utilize the English language equivalent as well as the acronym "NUB."

3. There are several specialized libraries on campus (e.g., business, agronomy) that offer updated resources. These are, however, limited as well.

4. See www.cnf.bj.refer.org.

5. Although our projects were completely independent of one another, we did invite a Belgian representative to attend and present at a workshop sponsored by the IDP.

6. The International Affairs Office coordinated the project and managed the budget and reporting.

BIBLIOGRAPHY

Drain-Jordan Library, West Virginia State University. 2002. Annual Report. Institute, WV.

Kistler, John M. 2003. "Special Acquisitions: Collecting African Materials." Pp. 31–50 in *Acquisitions in Different and Special Subject Areas,* edited by Abdulfazal M. Kabir. New York: Haworth Information Press.

Sharma, R. N., and Jeannie Bess. 2000. "West Virginia to West Africa and Back: An International Collaboration." *American Libraries* 31, no. 7 (August): 44–46.

"Strengthening Society by Building Skills with Public Archives: A West Virginia State [College] University–Université Nationale du Bénin Partnership." 1999–2002. The UNCFSP/USAID IDP Program. Quarterly and Annual Reports (1999–2002).

West Virginia State University Catalog. 2002. Institute, WV: University Press.

www.cnf.bj.refer.org (accessed August 5, 2004).

www.wvsc.edu (accessed August 5, 2004).

www.wvsc.edu/drainjordan (accessed August 7, 2004).

Chapter 11

The Impact of Information Technology on Africana Scholarship and Library Collections in the United States

Gregory A. Finnegan

ABSTRACT

Information technology (IT) facilitates scholarship and scholarly communication in African Studies in several ways. Electronic mail provides easy, and rapid, communications between scholars, between students and teachers, and between students, even when some are widely dispersed geographically. The Internet allows easy access to library catalogs, research resources such as bibliographies, and, not least, increasingly available full-text articles, books, documents, and manuscripts—many of them searchable for names and keywords in ways not possible before. Such worldwide access allows easier and more equitable participation in the scholarly enterprise, making it possible for scholars in more isolated places or countries to contribute more fully than previously, to the benefit of the knowledge we all create and share.

INTRODUCTION

African Studies presents several problems for academic and research libraries in the United States. The field—one cannot say "discipline" of an intellectual enterprise one of whose primary virtues is requiring its students to be cognizant of several traditional disciplines—is relatively young; the African Studies Association is not quite fifty years old. In addition, much relevant scholarship is published in small quantities, by small publishers, many of whom are outside well-established book-trade channels. Finally, African Studies is new not simply in proposing new approaches to the study of Africa but in pursuing such study at all. Because European and American cultures, and their scholars, paid little attention to a distant and "underdeveloped" continent, Africa was both strategically and literally marginal to all but a few academics. The long-lamented and persisting legacy of colonial transportation patterns, such that one had to fly to, say, Rome, to easily travel from Nairobi to Lagos, was mirrored in the difficulties of becoming aware of, and then acquiring, books, journals, and newspapers from Africa. As libraries have dealt with these issues, information technology has been a major factor in making support of scholarship deeper and more thorough, not to say easier. (Important views of African Studies libraries and Librarianship may be found in Witherell 1989, Schmidt 1998, and Evalds and Henige in press.)

INFORMATION TECHNOLOGY

General discussions of IT and scholarly research, communication, and education tend to focus on speed and convenience. The first emphasizes the great increases in speed of communication and dissemination of documents, as more and more scholars and students are linked to, and by, the Internet and the World Wide Web on it. The second stresses things like the ability to consult books, articles, and databases, and maps and images, from one's own desk, and with ability to quickly scan great swaths of text for names, concepts, or terms of interest. All this pertains to African Studies, even when not all such resources are as fully developed for our field as they might be for other disciplines. But an additional factor of critical importance to African Studies, if one still developing, is the extent to which electronic communication and access to electronic resources are leveling the playing field for scholars. Provision of infrastructure and greater network bandwidth, and even Internet access itself, are still issues being solved for Africa. Nonetheless, the ability of scholars in the so-called South to participate much more equally in international research and debate is the major consequence of contemporary IT (see Frank-Wilson forthcoming).

This would be true even if resources were more or less equitably distributed. The impact of worldwide use of jet airliners at the beginning of the 1960s did much to make mail and scholarly journals sent by mail more rapidly available to readers in Africa, Asia, and Oceania; the long time taken by ships to haul mail from New York or Southampton or Marseilles had meant scholars were months behind in their awareness of new ideas and developments in research and delayed their responses to others' writings as well.

The economic disparities between many or most African universities and their counterparts in the developed "North" means that more than just speeding up communication is important. The modern phenomenon of the "book famine" in African countries unable to appropriate foreign exchange to purchase new books and subscribe to journals, much less fill in gaps resulting from having newer institutions and inadequate purchasing power in the past, means that being able to provide electronic access to information, by Internet or even by CD-ROM or DVDs, is a significant step forward. Professor Raseroka has addressed the importance of such access within Africa in her chapter in this volume, but I want to emphasize here the equal importance for scholars in the North of easily receiving the research findings and opinions of colleagues from whom we've been for too long cut off by distance and weak means of travel and communication. The fact that Africans "at home" are increasingly able to participate in real-time discussion in forums like the "H-Africa" family of Listservs is equally valuable to all parties.

For reasons of relative development of communications and institutions, a most strikingly visible sign of how the end of apartheid opened up South Africa to full participation in the scholarly world was the presence of South African scholars and researchers in various Internet forums. Even before the rise of the World Wide Web, when "Gopher" technology dominated dissemination of information on the Internet, the African National Congress very quickly developed a very rich and deep online presence. I remember getting a reference question from someone seeking the Sesotho lyrics for the then-new South African national anthem. Ordinarily, that's the sort of thing one would find in a reference book. However, the question was from a major publisher of such books, trying to compile a reference book for the post-apartheid country—a true "chicken and egg" problem where information was needed to create the resource to provide the information. As it happened, though, the ANC had a link to documents from the Constitutional Assembly—so I was almost immediately able to supply to the questioner not merely the needed text but the text as provided directly by the official body responsible for it.

AFRICAN STUDIES AND UNITED STATES

The value of contemporary information technology for African Studies in lessening the differences between "center" and "periphery" is not limited to equalizing access to, and from, Africa itself. In the United States, at least, there are officially funded "Title VI" Area Studies centers, so named from the section of the National Defense Education Act that attempts to redress American deficiencies in science, language, and area-studies scholarship. Europe also has long-established centers where scholars and students can meet and exchange ideas. But the fact that, in the United States at least, a vast number of smaller colleges exist, with their emphasis on undergraduate teaching rather than research, means that many scholars work in places where they may be the sole Africanist, without fellow specialists to provide stimulation between annual meetings of the African Studies Association. Also, many small colleges are geographically distant from research centers of any sort and often are even farther from major centers for African Studies. Even within North America, the development of Internet-based resources, most notably the "H-Africa" family of Internet discussion lists fostered by Michigan State University, is a major change in making it possible for "Lonely Africanists," to use Corinne Nyquist and Leon Spencer's play on the title of Colin Turnbull's 1962 book to be in much more continual contact with other scholars and students (Nyquist and Spencer 1984). The publication just mentioned was published by the Archives-Libraries Committee of the (U.S.) African Studies Association. That group, now called the Africana Librarians Council, enjoys the distinction of being the oldest sponsored organization of the ASA, having been founded in 1957 at the same time as the association itself. This was not a random coincidence; it was apparent to the founders of the new interdisciplinary area-studies group that study of Africa in the United States required acquisition and cataloging of books, scholarly journals, and newspapers about, and from, Africa. The difficulties, and expense, of doing so meant that a few major institutions built the prominent collections of Africana: Northwestern University, above all, with its Melville J. Herskovits Library of African Studies, the Library of Congress (the major part of the de facto U.S. national library), as well as the universities of Indiana, Wisconsin, and Illinois, and Boston, Stanford, and Yale Universities, were in the first rank. Problems remained, however, about how best to ensure the most efficient acquisition of the widest range of materials and the most effective means of making scholars and students aware of new books and their locations.

Passage by Congress of the National Defense Education Act of 1958, a year after the founding of the ASA and the ALC, was spurred by the shock of the launch of Sputnik I, the first artificial earth satellite, by the USSR, in combination with a recurring awareness that the United States was failing to adequately understand the rest of the world with whom the country had to coexist. Title VI of the NDEA states that:

> The commissioner will contract with colleges and universities for the establishment of institutes to teach modern foreign languages if such instruction

is not readily available to individuals in government, business, or education. If understanding of a foreign region is necessary, these centers will also teach the history, economics, geography, and so on of the region.

An effect of the establishment of such Area Studies Centers at U.S. universities, including those for African Studies, was that it became more or less required that such universities have on staff specialist librarians to address the particular issues of acquiring library materials not often found in ordinary book-trade channels. Such specialist librarians were, and are, the core of the membership of the ALC.

Beginning in 1962, an early collective effort to address information needs of African Studies was Northwestern's coordination and publishing of *JALA*, the *Joint Acquisitions List of Africana*. This bimonthly serial contained catalog records for new acquisitions at the major Africana libraries. It served both to allow the major libraries to know of items they should seek to acquire or which they could forgo because others had them, while also allowing nonspecialist libraries and interested scholars to learn of new resources that might otherwise escape their attention.

JALA ceased publication in 1996, however, as something no longer needed because of modern information technology. A premodern information technology for libraries had been standardization of the three-by-five-inch catalog card and the sale by the Library of Congress, beginning in 1904, of printed card sets of its cataloging for new books, allowing other libraries to make use of the effort and expertise of LC to provide access for their own local collections. Beginning in the late 1960s, and becoming widespread by the end of the 1970s, however, North American (and increasingly, worldwide) libraries began sharing their cataloging by building enormous shared databases of catalog records, each "tagged" with a symbol indicating which libraries held the book or serial. Initially, these so-called bibliographic utilities required dedicated terminals wired to a national host computer. (As of the end of 2004, OCLC [see chapter 2], one of the two major utilities, held 57 million bibliographic records, representing holdings of 9,000 libraries.)

The existence of these huge bibliographic databases, eventually accessible over the Internet not merely to member libraries but also to the faculties, staffs, and students of their institutions (and recently to the world at large), meant that the current-awareness functions of resources like *JALA* were superseded by easily searchable databases. The rise of the Internet in the 1990s carried this change a step further, in that it's now possible in almost all cases to search the online catalog of specific libraries freely, from anywhere in the world, not merely to learn of resources but to plan visits to consult them. While progress still varies, more and more large research libraries have retrospectively converted their old card or book catalogs, including records in online catalogs; this means that bibliographic records for older as well as more recent materials may all be located online. As this development progressed, then, a paper product like *JALA* was no longer needed, and the information it contained was freely available to anyone, whether or not they or their library subscribed.

Library catalogs mostly index books and journals but rarely describe individual articles in them. For books, it is now easy for researchers to identify and locate information sources, regardless of where the researchers or the resources might be in the world. And, as noted, the very widespread access to e-mail in the developed world and increasing, if still less easy, access in the rest of the world, means that scholars can discuss new ideas and information directly, and, in online forums like H-Africa, as groups. There are, however, still other ways information technology enhances African Studies, as well as the potential for resources or ways of using them not yet developed.

Today's students, never having had to deal with difficulties in simply identifying or locating resources they might want, do not always share the awe that older researchers have for online bibliographic tools. They don't just want to know where something of potential use might be—they want access to such information, now, and here. And that in turn means they want direct access to the full text, not merely directions in the form of holdings information to follow to obtain a book or journal. Finally, modern students are not always as clear as they should be about what is a book and what is an article. All this, then, means that libraries feel pushed to deliver more and more full-text information, more and more easily, to readers.

A positive aspect of the demand for online access to full text is that, so far, the great proportion of what's available consists of articles from journals—not least because the initial drive for faster dissemination of information directly to users' desktops came from the "STM"—Science, Technology, and Medicine—sector, where journals are by far the dominant form of publication for research. This is significant in two ways for African Studies. One is that there is, as yet, no one dominant and authoritative online index to articles in African Studies, while there is more than one in print format. Online indexes relevant to African Studies certainly exist; examples include *Historical Abstracts* and *Anthropological Literature*, but an index to African Studies as such is lacking. To the extent, then, that online aggregations of journal articles exist that may be searched for keywords, names, or phrases, there is a partial equivalent of indexing where none existed before. Unifying subject indexing to ensure thorough coverage of a topic may still not exist, but the ability to find even incidental occurrences of key terms in bodies of texts is a great benefit. Second, and perhaps most important, is that vast numbers of articles are brought under bibliographic control.

In 1989, when there were few online indexes available to scholarly journals, and those few were both relatively shallow in time depth covered and accessible only through mediated searches performed by librarians, a writer noted that only 2 percent of the information held by research libraries was under bibliographic control. That is, because catalogs indicated only years, volumes, and issues held and the broad subjects covered by a particular journal, the individual authors and topics could be identified only by separate indexes, with significant additional work on the part of the would-be reader (Tyckoson 1989, quoted in Potter 1989).

In these circumstances, having some nineteen African Studies journals available as searchable text-images in JSTOR (for "journal storage") means a good percentage of the major English-language journals can now be searched. JSTOR is not a complete solution, since it is based on two significant compromises. Most important is avoidance of publishers' fears of loss of sales by unauthorized dissemination of digital information—most notorious with respect to music recordings and digital file sharing by means of programs like Morpheus. JSTOR therefore excludes the most recent years of journals, leaving sales of those to their publishers. In most cases, the rolling blackout of current issues is five years, but in some cases (like the journals of the American Anthropological Association) the period can be as much as seven years. JSTOR, then, is most suited for finding and accessing older information, rather than supporting current awareness. This is congruent with the origins of JSTOR as a means to reclaim shelf space by allowing long runs of journals to be moved from expensive library stack space to cheaper off-site storage, because their contents are available electronically.

The other major compromise of JSTOR is in the process of digitizing. Pages are scanned as PDF bit-mapped images—essentially electronic Xeroxcopies. Those in turn are run through an Optical Character Reading (OCR) program, one that JSTOR is persuaded is 95 percent accurate. The expensive step of proofreading is skipped, in the expectation that 95 percent accuracy is sufficient and that at least one instance of a word in an article will scan correctly. That scanned text is then loaded into the JSTOR search engine, underlying the JSTOR interface. A search then retrieves the PDF image associated with the word(s) searched; a user can then read, print, and save the text images but cannot copy text to paste into a document being written. Nonetheless, having access to long runs of key journals offers a kind of access previously available only by tedious review of indexes and tables of contents. In at least some disciplines, anecdotal reports indicate that currently valuable data and analysis, long forgotten, is being rediscovered by searches of JSTOR. Another virtue is that journals are not added randomly or piecemeal but are added only as part of disciplinary bundles judged to have "critical mass" for research in a given area.

Other aggregations of journals exist, though most or all of them lack the long runs, and may be perhaps better for current awareness. From the perspective of users, however, and, especially, librarians assisting them, a current problem is that relations between publishers and data aggregators are so volatile that one cannot assume that currently available titles will continue to be available in future. Harvard University, for example, maintains an online list of some ten thousand magazine and journal titles that have at least some content available online in full-text form but are not directly cataloged because of uncertain continuity; one must check the online catalog and the supplemental list to be sure that a given title and issue is or is not available online to our users.

It is also the case that online text, whether aggregated in useful groupings or as direct subscriptions, usually requires subscription fees, limiting access to those currently connected (as distinct from, say, alumni) to institutions willing and able to pay the necessary fees. An Africanist in a small college, for example, much less one in a country where foreign exchange is scarce, may often not have access to many of the resources mentioned here. There are possible developments that might change this. In the United States, there is increasing political pressure to ensure that publication of research funded by the federal government is freely available to citizens at large, whose taxes enabled the research to occur. This question is currently being raised with respect to medical research, but insofar as much academic research in the United States is made possible by government funds, this could eventually make Internet access much freer and wider, if perhaps constrained by (say) a six-month embargo during which only paid subscriptions could access the newest publications.

With respect to "the lonely Africanist" who teaches at a small institution that cannot afford subscriptions to resources not widely consulted locally, another possibility may be consortial access. As this paper was being written, for example, the American Anthropological Association was about to announce the first stage of an enterprise that is envisioned to allow the association's members online access to all the association's past and present journals and newsletters. This will include numbers of publications too narrow or minor to be included in resources like JSTOR, and it would allow members equal access, regardless of the size or location of their institutions.

Not all significant research appears in journal articles, of course. The various models of publishing (or, usually, digitally republishing) e-books are too varied, and too new, to say more than that efforts continue to find economical means to market enough books in electronic form to make it worthwhile for libraries to invest in them, assured that reader access will be easy and that access will continue to exist—especially without open-ended payments to continue it. Yet another form of publication of the newest findings on a topic is the doctoral dissertation. Since a PhD or equivalent degree certifies originality of scholarship, being either new information or new analysis of older information (which thereby becomes new information), dissertations must be available to other scholars to evaluate and benefit from. In the past, this meant printing or otherwise duplicating multiple copies, to be exchanged with other universities for their dissertations. Beginning in the 1950s in North America, a private company, University Microfilms, assumed for almost all U.S. and Canadian universities the task of microfilming and maintaining an archive of dissertations, in exchange for exclusive rights to sell microfilm and paper copies.

In recent years, this process has increasingly included dissertations from universities outside the two originally participating countries, though world coverage is far from complete. Since the end of the 1990s, in addition, dissertations are accepted in digital form and are digitized upon receipt if submitted traditionally. While done in part because dissertations themselves increasingly included nonprint components—video and sound recordings, for example—and because making only the print portions available limited the utility of the (now partial) dissertation for others, digitization of dissertations also makes delivery of them to purchasers very much easier and quicker. Indeed, dissertations from past decades are now routinely digitized whenever a copy is ordered, so the proportion digitally available continues to grow. Although individual scholars can now purchase digital copies of dissertations, as they previously could in paper or microform, institutions can also subscribe to the service, so that their students and faculty can freely download dissertations. As with so many resources described here, costs are not insignificant and not all institutions can afford all that their members might desire. Nonetheless, dissemination of original research in dissertations is now easier and faster than it was before easy digitization and the Internet were available.

Another developing facet of dissertation exchange is the availability of dissertations from the rest of the world—in our case, most of all, those from African universities—to scholars elsewhere. This is a complex matter, especially because less-developed nations have long experienced having resources, material as well as intellectual, being exported to the greater benefit of other nations. Insofar as the politics of information exchange are resolved, available IT does mean that actual transfer of information will be easily accomplished.

Earlier tools like *JALA* were fundamental to the necessary awareness of new publications required before libraries could acquire them. Publishers have long issued printed catalogs, of course, and various national bibliographies have documented publications in their countries since the nineteenth century. Those bibliographies often lag in their coverage, however, and have been especially irregular from less developed countries. Mailing expenses aside, publishers and libraries need to know of each other's existence (and addresses) before catalogs can even be requested. The Internet and e-mail have made it significantly easier for publishers and librarians to be aware of each other and to communicate easily, cheaply, and quickly.

It is now common for publishers, including the small and/or relatively isolated ones that play such an important role in Africana publishing, to announce new books to potential buyers. (Some of these announcements, along with tables of contents of new issues of journals, appear on resources like H-Africa, alerting potential readers as well as individual buyers to the availability of new works.) Thanks to efforts like those of the African Books Collective, the organization owned by seventy-four African publishers that works to make African imprints easily known and available in mainstream book-trade channels, it has become easier for more potential buyers to know of relevant titles. This not only improves the financial situation for publishers, but sales in Europe and North America (and Japan) mean sales that generate foreign exchange needed by publishers (and their home countries), further supporting an essential industry.

Information technology has also opened up the infrastructure of libraries themselves to wider participation. The Cataloging Committee of the Africana Librarians Council, for example, operates a "funnel project" in which specialist Africana catalogers are able to establish new subject headings needed for new books—headings acceptable to the Library of Congress as administrator of the dominant subject-heading system used in English-speaking libraries. Before this change, only the Library of Congress itself could establish new headings; the wider participation is effectively possible because of e-mail communication between catalogers who might otherwise meet only twice a year, and because necessary supporting documentation can be easily submitted to LC electronically.

There are other areas of African Studies affected by information technology. One is the role of computers as means of recording and analyzing research, including research in the field. As far back as 1980, I remember a Purdue University–administered research project concerning agricultural development in what is now Burkina Faso. The project used Apple-II computers because their circuit boards could easily be mailed when repairs were necessary. Technology has clearly improved in the almost quarter-century since, with the result that data can be recorded and analyzed much more quickly. Twenty years ago, Professor Joel Samoff of

Stanford University was the discussant on a panel at the African Studies Association meetings on "Computers and Africanist Research" (October 27, 1984). In his presentation, Professor Samoff mentioned work he had done in teaching; not wanting to spoon-feed a reading list to his students, but also not wanting them to sink out of sight in the new online library catalog, he downloaded two or three hundred records. This was enough to force intelligent choice of resources by his students, but from a bounded universe of material considered valid by the professor. In that mainframe-computing era, he then uploaded the selection of records to a file for his students to access—with the additional bonus (as he saw it) of being able to reconfigure the records to a format he preferred to that used by the library. We're still not in a situation where users (or teachers) can routinely customize catalogs and catalog records, but clearly we will move more and more in the direction of allowing users to not merely access information but to do so in individually configured ways. Online discussion lists for students in classes are also increasingly common, allowing easy communication between them but also allowing professors and specialist librarians to intervene as well—suggesting, for example, new information resources relevant to a new topic someone has raised.

Librarians still envision a new generation of catalogs more than they implement them, but some work is suggestive of what might come. Janet Stanley, librarian of the Warren M. Robbins Library of the National Museum of African Art at the Smithsonian Institution, editor of the important serial *The Arts of Africa: An Annotated Bibliography*, has enhanced catalog records for her library so that, for example, the record for a book also contains citations to reviews of that book—saving the subsequent effort to find scholarly evaluation of the book. As research encompasses more and more formats, even the concept of a single library catalog is changing. Harvard University, for example, has an online catalog of traditional records for books, serials, and other items. But it also has links to "other catalogs" such as specialized databases for visual images, for detailed finding aids for archival collections, and for geo-spatial information. No one format of catalog will accommodate the varied information provided by these different sorts of resources.

Another topic, too new to have had much impact on African Studies, is "distance education," providing class lectures, readings, and supporting materials to students unable to be present on a campus, by means of Internet access and e-mail. A leader in such work is the University of Iowa, which, under the leadership of their late African-specialist librarian John Bruce Howell, created in association with the ALC the *Electronic Journal of Africana Bibliography*. This online resource, freely available to all, at the end of 2004 contained nine peer-reviewed topical bibliographies on various African Studies topics.

CONCLUSION

Although IT may be less prevalent or prominent for African Studies (in contrast with scientific, technical, and medical fields) it is nonetheless true that research itself, and the institutions and library collections necessary for research, could not work as effectively and efficiently as they do with it.

BIBLIOGRAPHY

Evalds, Victoria K., and David P. Henige, eds. In press. *African Studies Librarianship in an Era of Change: A Festschrift in Memory of Dan Britz*. Lanham, MD: Scarecrow Press.

Frank-Wilson, Marion. forthcoming. "Electronic Publishing and African Studies: A Way to Bridge the Information Gap?" In *African Studies Librarianship in an Era of Change: A Festschrift in Memory of Dan Britz*, edited by V. K. Evalds and D. P. Henige. Lanham, MD: Scarecrow Press.

Nyquist, Corinne, and Leon P. Spencer. 1984. *The Lonely Africanist: A Guide to Selected U.S. Africana Libraries for Researchers*. Los Angeles: Archives-Libraries Committee, African Studies Association.

Potter, William Gray. 1989. "Expanding the Online Catalog." *Information Technology and Libraries* 8, no. 2: 99–104.

Schmidt, Nancy J., ed. 1998. *Africana Librarianship in the 21st Century: Treasuring the Past and Building the Future: Proceedings of the 40th Anniversary Conference of the Africana Librarians Council*. Vol. 6, *Monographs on Africana Librarianship*. Bloomington: African Studies Program, Indiana University..

Tyckoson, David. 1989. "The 98% Solution: The Failure of the Catalog and the Role of the Electronic Database." *Technicalities* 9, no. 2: 8–12.

Witherell, Julian W., ed. 1989. *Africana Resources and Collections: Three Decades of Development and Achievement: A Festschrift in Honor of Hans Panofsky*. Lanham MD: Scarecrow Press.

IV

MIDDLE EAST

Chapter 12

The Impact of Technology on Libraries and Collections in the Arab Countries of the Middle East and North Africa

Mohammed M. Aman

ABSTRACT

Information technology (IT) has impacted the operations and services of libraries worldwide as well as users' expectations and aspirations. In the Arab countries of the Middle East, change has been painfully slow due to a number of socioeconomic, political, and technological factors. In this chapter, I will discuss these factors and point to the latest statistics from the United Nations and other international organizations on the levels of technology use in these countries. I will also describe some of the accomplishments of Arab countries in introducing IT in their libraries and information systems and the emergence of an IT business and industry in these countries that can lead to further unrestricted access to global information exchange. I will suggest that these phenomena promote a fledging democracy, free-market economy, and, it is hoped, some economic and political stability in that troubled part of the world.

INTRODUCTION

Faced with rapidly growing populations and a declining gross national product (GNP), the Arab countries are eager to modernize in a world where use of computers and computing technologies has become indispensable. The ability to connect with people from all over the world expands the idea of a global knowledge-based society and enhances understanding of the diversity of customs, beliefs, and cultures in the world. The dynamics of interactions, Internet, electronic mail, and video conferencing have transformed our assumptions about communicative norms. The expanding capacity of the cellular phone has catapulted even the most remote parts of the Arab world into the global communication network. Electronic-based interactions now sustain and support many important dialogues in national and international settings. Any part of the world, including the Arab world, that is not accessing these technologies will not be an active participant in the global development of world affairs in the twenty-first century. Even as access to computer technologies has the potential to empower local stakeholders and support the democratic development of an information-literate society, active participation in the Internet community sustains an informed public.

THE ARAB SITUATION

The "Arab world" refers to the twenty-one countries that are members of the Arab League. They share a common language and a predominant religion. Arabic is the official language, and Islam, with the exception of Lebanon, is the dominant religion in these countries. They also share common history, religion, customs, and traditions. Except for a handful of oil-producing countries, almost all of the Arab countries have low per capita income. Even among the oil-producing countries, known as "rentier states," youth unemployment in many cases is running at between 30 and 50 percent. The need to replace foreign workers with locals, as in Saudi Arabia, raises difficult issues of labor productivity and income distribution. Population growth and lack of national planning have constituted obstacles to moderate quality of life in the region. In 1970, the six oil exporters on the Gulf—Iran, Iraq, Kuwait, Qatar, Saudi Arabia, and the United Arab Emirates—had a total population of about 45 million. By 2000, this figure had more than doubled to about 117 million. Economic growth rates have been lower than population growth rates, so that average incomes have fallen. Population growth and uncertain oil prices mean that per capita oil revenues in the Middle East are far more likely to fall than to rise (Heradsveit and Hveem 2004).

Unemployment has risen, especially among young people, as governments no longer have resources to both educate and employ them. The result is a higher risk of social distress and political instability, as we now see events unfolding in Saudi Arabia.

The economic monoculture of oil has made the Middle East totally dependent on that single commodity market. High revenues in the past have not led to any diversification of the economic basis, except in small rentier economies such as Kuwait and the United Arab Emirates, where a substantial part of the economic surplus has financed foreign investment. Since the mid-1980s, low oil prices have forced severe cuts in public budgets. One example is Saudi Arabia in 1994, 1995, and again in 1998. Budget constraints have led to transfer cuts and declining living standards for large parts of the population.

Education

Education in the Middle East is facing a critical challenge to meet the demands for the twenty-first century, with its ever-increasing population and declining GNP. This means that those seeking access to education at all levels—primary, secondary, and tertiary—will increase. During the 1970s and 1980s expenditure on education had increased drastically in all oil-exporting countries. Today, there is modest expansion to accommodate the increasing number of students who are seeking access to higher education. In addition to expansion of old and well-established universities, new public and private universities are being established. Examples are the American University in Kuwait (AUK), the Abu Dhabi University in the United Arab Emirates, the Sixth of October University in Egypt, and the Open University in Egypt and Lebanon, among others. Alternative ways of providing access to higher education via distance education are also being explored in a number of Arab countries in the Middle East (Aman 2004).

The Arab countries of the Middle East and North Africa have a large number of young people who are college educated and willing to do anything for a living but who unfortunately cannot find employment in either the private sector or in the already overcrowded public sector (Ende 1991). This circumstance has become increasingly acute in Egypt, Morocco, Kuwait, and Saudi Arabia, in addition to Iran, whose young people often have high expectations. Governments in the oil-producing countries of the Middle East and North African Countries (MENA) have not dared to impose income and property taxes, with the result that they are welfare states, unable to cope with population growth, a monoculture economy, and low self-sufficiency for food supplies.

Equally distressing is the lack of reform in the Arab education system, which seems to be stuck in the nineteenth century. The system either relies on rote learning and lecture notes or is single-textbook-based, with no requirements for outside reading or independent research or thinking. Therefore, students are absent from Arab libraries, except for those who love to read regardless of what teachers teach. Arab libraries do not have the reserve collections, e-reserve, or even the reference services or desks that we are accustomed to in the United States.

Arab and North African countries have high levels of illiteracy, hovering around 50 percent. Internet access means nothing to someone who has not learned how to read or write. The United Nations has designated ten years of this century as the Decade of Literacy (from 2002 to 2012). Even among educated college graduates, poor command of the English language makes use of the predominantly English-language Internet hard to use.

Until very recently most Arab universities were excluded from the information society. Exceptions can be found at the American University in Cairo and its counterpart in Beirut. Both are modeled after the American system of higher education, and both are financially out of reach for children of the average Arab family. The knowledge economy requires populations with training of an even higher standard in the interest of enhancing productivity and maintaining a competitive edge.

Censorship

The advent of the Internet in the Arab countries of the Middle East has challenged the prevailing strict, government-imposed censorship. Article 19 of the Universal Declaration of Human Rights is unambiguous in its stance against any form of censorship. To put it into practice in most, if not all, Arab countries is more difficult than declaring it. Such censorship and the overzealousness of government officials about protecting the citizens from outside influence have placed severe barriers to the rapid expansion of the Internet and access to the global information marketplace. The fear of cultural invasion from the Internet is more predominant in the thinking of government censors than what is being shown on satellite channels, as one can judge by the satellites growing like mushrooms in balconies and on roof tops of the most modest of Arab homes.

Political Instability

In the Middle East, political stability requires high oil revenues and gradual development of representative institutions. On this latter point, Iran is the most advanced, Iraq the least. The Islamic movements in the Arab Muslim countries would appear to represent a social revolt as well as an assertion of cultural and national identity in the wake of incomplete modernization. The present autocratic systems of government do not facilitate dialog and compromise but provoke conflict instead. The outcome is intense social and generational tensions. In most Arab countries the established political leadership is not accountable to the population through democratic processes. The social crisis leads to a cultural crisis with religious references and political significance. It prepares the ground for Islamic movements.

Poor Publishing Output

No more than ten thousand books were translated into Arabic over the entire past millennium, equivalent to the number translated into Spanish each year. Egypt and Lebanon are the leading publishing centers of the Arab world.

To illustrate how poor the publishing industry is in the Arab world, let me cite some comparisons. The entire Arab world produces about four thousand books annually; most of them are reprints or religious books (*turath*, or cultural heritage). By contrast, Holland, with a population of less than 4 percent of the population of the twenty-one Arab countries, produces 44,000 books a year.

Scientific Research and Publication

Government spending on scientific research is no better than it is in the areas mentioned above. No Arab country spends more than 0.2 percent of its gross national product on scientific research, and most of that money goes toward salaries. By contrast the United States spends ten times that amount; fewer than one in twenty Arab university students pursues scientific disciplines. There are only eighteen computers per thousand people in the Arab world. The global average is seventy-eight per thousand. Only 370 industrial patents were issued to people in the Arab world between 1980 and 2000. In South Korea that same period, 16,000 industrial patents were issued (Castillo 2004).

Absence of Information Policies or Strategies

A prerequisite for national information development and infrastructure is a national information policy for any or all of the Arab countries. The absence of clearly defined national information policies in almost all Arab countries, perhaps with the exception of Israel and Iran, poses another challenge. Information policies are needed to provide a framework for the development of computer-based information and library systems in the countries of the region. An information policy could guide their development in these countries. The absence of such a policy is a clear obstacle to the development of Computer-Base Library and Information System (CB-LIS). The presence of such a policy or strategy could result in the elimination of duplication and better coordination of information resources and services within a country.

Attempts were made in the mid-1980s to establish an Arab information network. The main effort was spearheaded by the Arab League Documentation Center (ALDOC), which was established and later expanded in

Tunis during the 1980s. One of the projects of ALDOC was the establishment of ARISNET (the Arab Regional Information System Network), which had among its goals the establishment of an Arab Information Network. In Kuwait, the Kuwait Foundation for the Advancement of Science (KFAS) is active in supporting research. In the UAE, a national research foundation was established to pump money into scientific research and help establish research-based doctoral programs at UAE universities.

Efforts are under way to establish a Gulf-based Arab Science Foundation at Sharjah (like the American NSF and KFAS in Kuwait). It will grant $10 million in grants to Arab scientists to conduct research every year. Specialized private universities are spreading throughout the region. Some are looking for national accreditation by the same bodies that accredit academic programs in the United States. There is recognition in some Arab countries that scientific research can fuel economic development and growth.

THE LIBRARY SITUATION IN THE ARAB WORLD

The literature on the library situation in the Arab world shows that libraries there suffer from a number of ills, including poor collections due to lack of public funds, poor local publishing output, and poor support from the government agencies that fund these libraries and their parent institutions. There are major gaps in serial holdings, especially scientific journals. Cooperation among libraries within the same country is almost nonexistent, not to mention cooperation across these countries. Equally lacking is interlibrary loan, cooperative acquisition, and the like. In the areas of collection development, librarians play minor roles in selecting materials and faculty are ignorant of bibliographic sources on the basis of which they can make educated judgments about selecting quality materials for their college libraries.

As we observe library expenditures, we are struck with the fact that money, when available, is being spent on buying expensive equipment and hardware rather than on books and journals. In most cases, the selection of hardware and software is not well thought out, and plans for continued maintenance and upgrades are usually nonexistent, or not taken into account for future budget planning.

Authorities are often technologically illiterate and blindly convinced that automation is a panacea for all library ills. They insist on buying technology for the wrong reasons. Among the reasons cited for automations are to be fashionable (faddishness), to look like a leader in IT, or to make money for selfish, unethical reasons (financial profiteering or kickbacks).

To the professional observer, Arab libraries require a great deal of work before computerization can become a viable priority. Examples of needs for improvement include better inventory control, weeding or deselecting obsolete materials, book repairs and preservation, better cataloging, better professional support to clients, and more innovative user-oriented services.

THE DIGITAL DIVIDE IN THE ARAB WORLD

The digital divide, as Adam Samassekou, president of the World Summit on the Information Society Preparatory Committee and president of the African Academy of Languages, puts it, "is but the most visible aspect of a set of more serious and deeper divides that warrant the concern of the international community." The information society, he added "is not characterized by the availability of technologies, but is a body of economic, cultural, social and political phenomena which together define a new stage in the history of humanity" (Samassekou 2004).

Arabs have limited access to the information highway either at their libraries or at their homes. There are many homes without phones, and in rural or desert areas, without electricity. All too often, technological inputs are received as imported commodities that are not linked to the existing social, cultural, and material sources of a region or country. The introduction of technology occurs without a thorough understanding of the interrelationships needed to implement, manage, and sustain its infrastructure The gap between urban and rural communities is ever widening, as is the gap between the haves and have-nots. The latter constitute 25 percent of the population.

ARAB ADVANCEMENTS IN INFORMATION TECHNOLOGY

In addition to the societal, political, economic, and educational constraints mentioned above there are a number of technological constraints that hinder the expansion of computerized information systems in the Middle East. Paramount among them are telecommunication and other communication infrastructures. Connectivity beyond major and capital cities poses a potential problem in creating a national information strategy. Library and information users would need access to computers that can send and receive messages using browsers such as Explorer or Netscape. In addition, they would have to find on their computers word processors and other applications to make full use of databases and access to virtual libraries from around the world. Easy and inexpensive connections to Internet service providers would be required if Arab students and scholars were to have access to their local or global libraries and information systems. Clearly related to connectivity issues are financial matters. Another challenge is the lack of trained cadre for professionals to support the implementation of computer-based library/information systems. Few Arab librarians have expertise in implementing these systems. There is a great deal of reliance on system vendors, many of whom do not maintain local offices in the areas where their systems have been installed (Misik 1993).

Computer-based library and information systems, by their very nature, involve more than just the transmission of cultural/social paradigms between and among the participants. Libraries and information systems do not operate or flourish in a vacuum. There have to exist external forces and reasons for using information, alike for individuals and for national development. This will require a new paradigm shift in the way educational development and planning are viewed by the government authorities and decision makers.

Commonly Known IT Systems and Services in the Region

A private Saudi company, Nuzum al-Ma'lumat al-Mutatwirah, has introduced an Arabized version of the library software Horizon, which is marketed under the Arabic name al-Afaq. The system is now used in a number of academic libraries throughout the region.

Another company located in Palestine known as NourSoft markets its LibSys, which is an integrated Arab software for all sizes of libraries, information centers, and archives (www.libsys.net).

Other more familiar systems, like VTLS and Innovative, are also been Arabized and are being used in a number of libraries, including the new Library of Alexandria, Abu Dhabi University Libraries, the Saudi Documentation Center for Business & Finance, KISR's NSTIC, and others.

CDS/ISIS is being used in a number of small Arab libraries belonging to Arab League organizations or ministerial and government libraries and information centers throughout the Arab world.

A number of electronic libraries and databases have emerged in the Arab world. Among them is Al-Awraq (www.alwaraq.com), which is an Arab electronic library containing nearly five hundred sources of information on Arab heritage (mostly materials on religion and Arab classics).

Most major Arab newspapers are now available online. Examples are:

Al-Ahram (Egypt) (www.ahram.org.eg)
Al-Ayam (Bahrain) (www.alayam.com)
Al-Gumhuriyah (Egypt) (www.tahrir.net/algomhuria)
Al-Hayat (London-based) (www.alhayat.org)

Al-Ittihad (UAE) (www.alittihad.co.ae)
Al-Jazeera (Qatar) (http://english.aljazeera.net)
Al-Khaleej (UAE) (www.alkhaleej.co.ae)
Al-Qabas (Kuwait) (www.alqabas.com.kw); *Arab Times* (Kuwait) (www
  .arabtimes.com)
Al-Watan (Kuwait) (www.alwatan.com.kw).

Arab search engines include www.Kuwaitview.com, which provides information on many aspects of life in the Arab world such as women's rights, peace, jobs, etc.

CAN IT AND LIBRARIES HELP IN USHERING DEMOCRACY INTO THE ARAB WORLD?

With talk increasing in Washington, London, and other Western capitals and in the Middle East about the necessity of introducing political and economic reform in the Arab and Muslim world, we must help facilitate the role of Arab libraries in introducing and supporting these reforms. The new Middle East Initiative introduced by the American administration has created hopes among the masses and anxiety among the rulers in the Arab Muslim world. The Middle East Initiative calls for establishing civil societies and democracies, two important pillars on which the American library system has been founded. It will be very appropriate for the American Library Association to help Arab librarians in accomplishing the tasks outlined in the Middle East Initiative without appearing to be imposing or dictating an American will or engaging in a cultural invasion.

The new Bibliotheca Alexandrina (BA) has claimed a place for itself in the history of the Arab movement toward freedom, democratization, and civil society. It was at the BA that the first conference on Arab reform issues was held from March 12 to 14, 2004. The conferees agreed, among other measures, to establish an Arab Reform Forum at the Bibliotheca Alexandrina to act as an open forum for initiatives, intellectual dialogue, and Arab projects. With regard to cultural reform, the conferees agreed, among other measures, to provide a cultural atmosphere to promote democratic development and peaceful transfer of power. This, the conferees added, "can only be achieved by confronting political systems that can prevent any effective political participation or reform" (*Alexandria Statement* 2004, 24). They also called for "reforming and activating Arab cultural institutions through financial and moral support to widen the range of their plans and coordinate between them and other cultural organizations" (24).

Arab librarians can learn from their American colleagues how to involve the local citizenry in the governance and running of their local public library—forming something akin to what we have in the form of a public library board of trustees. Arab public libraries can also help in forming debate societies, book clubs, book discussion groups, and Internet groups that can be nuclei of civil societies for young people. This will keep them off the street, out of trouble, and, it is hoped, retooling their education and skills to fit the requirements of the job market. Arab public libraries can assist in promoting computer and information technology literacy among the masses. Access to the world information sources is essential for informed citizenry in this part of the world. Laptop loan programs for young and old and free access to the Internet in local public libraries can assist in the efforts to establish democracy and free-market economy in this part of the world.

Part of the Middle East Initiative is also education reform, which is desperately needed in the Arab world. As I have mentioned before, there is a need to move away from the textbook-based method of education to the more library-oriented method of teaching, which relies on outside readings and directed research instead of rote learning and memorization for the sake of passing final examinations.

Private universities are emerging, and this is an opportune time to introduce innovative systems of learning in these schools with emphasis on entrepreneurship, globalization, science and technology, including IT, and knowledge of foreign languages.

American publishers can assist with reforming and energizing the stagnant Arab publishing industry and adding variety to what is being published, and they could be more responsive to the needs and desires of the Arab reader. Quality books for children and young readers can help in promoting love of reading and library use among young people, the future generation of Arab readers and library users.

CONCLUSION

Computer-based information systems and libraries without walls should be appearing in the Arab countries of the Middle East. There are many obstacles and constraints on the road to implementation; paramount among them are educational and economic hindrances. The complexity in introducing computer-based systems in the Arab world is enormous. In spite of the challenges confronting the advancement of computer-based information systems and technologies in the Middle East, there is a growing interest in the concept. The initiative of education reform represents hopes for millions of Arabs looking for access to global libraries and information systems. Arab librarians and information professionals should take advantage of professional opportunities, gain access to up-to-date materials, provide virtual and onsite access to their library/information users and others around the world, and become part of the global information planning community.

Arab library and information science associations are increasingly aware of the potential of computer-based information systems in addressing national development plans. The activities of these associations such as AFLI (Arab Federation of Library Associations) (www.afli.org) in Tunis, Cybrarians (www.cybrarians.info), and Librariannet (www .Librariannet.com) in Egypt and others have contributed to this awareness. These and other associations are having a tremendous impact on shaping the future. Arab states should establish low-cost and self-sustained access points to provide access to the Internet in all Arab countries. Each country should attempt to build information and communication infrastructure. Arab institutions must establish links to foreign partners utilizing virtual library cooperatives. Such programs will offer Arab library users the opportunity to take advantage of global library and information systems and databases.

I conclude this chapter by citing from the *Alexandria Declaration*:

> Information and data are extremely important to make decisions and for a realistic and sound analysis. There is a need to pass laws that would obligate authorities producing economic data to make this data available and easily accessible, wherever needed according to clear and agreed rules. Comprehensive databases of Arab economies should be prepared.

Access to information is an essential ingredient not only for and in a free-market economy, but also for all other sectors in a democratic and free society. May democracy, free and unrestricted access to information, and libraries prevail in a new Arab world!

BIBLIOGRAPHY

Alexandria Statement, March Statement: Final Statement of "Arab Reform Issues: Vision and Implementation." March 12–14, 2004. Alexandria: Bibliotheca Alexandrina in Cooperation with Arab Academy for Sciences & Technology, Arab Business Council, Arab Women's Organization, Economic Research Forum, and Arab Organization for Human Rights.

Aman, Mohammed M. 2001. "The Bibliotheca Alexandrina," Pp.1–10 in *International Librarianship: Cooperation and Collaboration*, edited by Frances L. Carroll and John F. Harvey. Lanham, MD: Scarecrow Press.

———. "Libraries in the Middle East: An Overview." 1994. Pp.1–1 in *Information and Libraries in the Arab World*, edited by Michael Wise and Anthony Olden. London: Library Association.

———. "Globalization of Distance Education." 2004. Paper presented at the 2nd International Education Conference, Honolulu, Hawaii, January 2004.

Castillo, Daniel D. 2004. "The Arab World's Scientific Desert. . . ." *The Chronicle of Higher Education*, March 5, 2004, pp. A36–A38.

Ende, Werner. 1991."Auf der Suche nach der idealen Geselschaft." Pp. 64–72 in *Die Golfregion in der Weltpolitik*, edited by Peter Pawelka, Isabella Pfaff, and Hans-Georg Wehling. Stuttgart: Verlag W. Kohlhammer.

Heradsveit, Daniel, and Helge Hveem, eds. 2004. *Oil in the Gulf: Obstacles to Democracy and Development*. Burlington, VT: Ashgate Publishing

Ismael, Jacqueline. 1993. *Kuwait: Dependency and Class in a Rentier State*. Gainesville: University of Florida.

Meho, Lokman, and Mona A. Nsouli. 1999. *Libraries and Information in the Arab World*. Westport, CT: Greenwood Press.

Misik, Abdel Halim. 1993. *Education on Information Specialists in the Arab Region*. Paris: UNESCO.

Noreng, Oystein. 2004. "The Predicament of the Gulf Rentier State." Pp. 9–40 in *Oil in the Gulf: Obstacles to Democracy and Development*, edited by Daniel Heradsveit and Helge Hveem. Burlington, VT: Ashgate Publishing.

OECD. 2001. *Understanding the Digital Divide*.

Samassekou, Adam. 2004. "World Summit on the Information Society: The First Step Towards a Genuine Shared Knowledge Society." *IFLA Journal* 30, no. 1: 6.

UNESCO. 1996. *Information and Communication Technologies in Development: A UNESCO Perspective*. Prepared by the UNESCO Secretariat. www.unesco.org/webworld/telematics/uncstd.htm#Education (accessed August 11, 2004).

Chapter 13

Information Technology Applications in Information Work in Egypt: A Puzzle Missing Some Pieces

Sherif Kamel Shaheen

ABSTRACT

This chapter is divided into six sections. It starts with describing Egyptian efforts to build an information society, and then, in the second section, it turns to presenting different Egyptian academic departments and institutes teaching multiple aspects of information technology (IT) applications in information work. The third section deals with IT in Egyptian library and information science research. The fourth section presents integrated library systems, digital library projects, and Egyptian library websites. The fifth section discusses the IT private sector in Egypt. The chapter ends with some important conclusions that are considered as missing pieces in the IT puzzle in Egypt.

THE EGYPTIAN WAY TO INFORMATION SOCIETY

Egyptian Government Plans for an Information Society

"Developing Egypt into an information society is a top priority." With these key words, Egypt's President Mubarak gave impetus to the development of information and communication technologies (ICTs) in Egypt, linking them to the economic and social development of the country. "The government of Egypt as a major stakeholder in the global information society is committed to building an Egyptian information society, offering every individual, business, and community the opportunity to harness the benefits of the new information era to achieve national priorities."[1]

The Cabinet—Information and Decision Support Center

The Information and Decision Support Center (IDSC) was established by the Egyptian government in 1985 to build up Egypt's IT industry and governmental decision-support infrastructure. IDSC evolved from the government's commitment to join the international IT revolution. One of its primary objectives was to provide public access to information, with a particular emphasis on aiding business and investment. Over the past years, IDSC has successfully implemented many IT projects in legislative reform, public-sector reform, human resources development and job creation, access to the Internet, commercial registration, natural resources management, cultural heritage preservation, urban planning, and sectoral development projects at the ministerial and governorates level, among many other areas. IDSC is currently focusing on decision support for the Cabinet of the Egyptian Information Society.[2]

The IDSC has been at the forefront of efforts to promote the use and institutionalization of information technology in the government. It has worked with the majority of Egyptian ministries to establish information and decision-support centers in the ministries and create decision-support systems for key sectoral issues, such as the customs tariff and the electronic tariff. It has also created hundreds of decision-support centers in all the Egyptian governorates (Egypt's administrative divisions).

Egypt's Information Highway Project

In order to mobilize and empower the development of the Egyptian information content on the Internet, IDSC and the Regional Information Technology and Software Engineering Center (RITSEC) have jointly launched Egypt's Information Highway Project, which is a pilot project that aims at supporting Egypt's socioeconomic growth. An umbrella project within which several subprojects have been initiated to tackle different crucial sectors, it involves culture, tourism, health care, environment, industry, trade, investment, local administrative divisions (governorates), and public services. The objectives of the project are:

- To promote and support electronic dissemination of information over high-speed communication networks (information highways);
- To establish pilot information highways in critical areas to energize socioeconomic development;

- To contribute to open and wide access to the national information highway;
- To encourage and support the development of secure online databases; and
- To assist in the human resource development required for establishing the national information highway.

The implementation of the project involved cooperation and coordination with other national initiatives, as well as establishing partnerships with government and private-sector entities.

Since the launching of the project late in 1995, several pilot information networks have been launched covering culture, tourism, health care, environment, education, public services, and governorates. Activities include content creation (online databases and home pages), human resource development, and creating and supporting user groups. Target groups include: investors, developers, health-care professionals, environmentalists, government officials, and the general public. Within this project, several pilot networks are being launched:[3]

Egypt's TourismNet provides basic information on Egyptian hotels, restaurants, cruise lines, travel agents, transportation companies, and tourist attractions. It contains several search engines that facilitate searching through tourism databases.

Egypt's CultureNet provides information of the Egyptian cultural heritage, arts, historical sites, and museums.

Egypt's HealthNet contains information on the Egyptian medical centers, physicians, medical companies, and medical laboratories. A search engine is provided for searching the physicians'database.

Egypt's GovernoratesNet provides basic statistical information on Egypt's administrative divisions (governorates).

Egypt State Information Service

The State Information Service (SIS) is the public information organ of the government of the Arab Republic of Egypt. Well-known counterparts are USIS of the United States, Public Information Office of Britain, and the Bundespresse Amt of Germany, among others. SIS was established in 1954 under the name of Information Department (ID) essentially to contribute to political socialization and to build a worldwide image of the newly founded republic in the wake of the 1952 Revolution.[4]

Ministry of Communications and Information Technology

In September 1999, President Mubarak announced the inauguration of a national program for the development of the communication and information technology sector. The national program goals were to create in Egypt the Information Society and an export-oriented ICT industry.

In October 1999, a new Ministry of Communications and Information Technology (MCIT) was formed to facilitate Egypt's transition into the global Information Society. The new ministry began its work by preparing the National Plan for Communications and Information Technology. MCIT's projects are geared toward supporting and empowering the Information Society in Egypt in close coordination with relevant government agencies and with the private sector. These commitments have been translated into developing and expanding the telecommunications infrastructure, establishing hundreds of IT clubs, expanding the pool of IT skilled labor, and creating national information systems and databases.

The National Plan for Communications and Information Technology has paved the way for the initiation of the Egyptian Information Society Initiative (EISI), which has been structured around seven major related tracks, each designed, when fully implemented, to help bridge the digital divide and facilitate Egypt's evolution into an Information Society. [5]

The ministry's six specific tasks under this initiative are:

E-Readiness: Equal Access for All. The Information Society should enable all citizens to have easy and affordable access to the opportunities offered by new technologies.

E Learning: Nurturing Human Capital. ICT is a complementary tool for higher standards of education at all levels and for upgrading the skills and productivity of the citizenry.

E-Government: Government Now Delivers. The Information Society should be able to deliver high-quality government services to the public where they are and in the format that suits them.

E-Business: A New Way of Doing Business. ICT is an important tool for robust economic growth.

E-Health: Increasing Health Services Availability. The application of ICT in the health sector could provide a better quality of life to the citizens and a more efficient work environment for physicians and health-care workers.

E-Culture: Promoting Egyptian Culture. ICT is used to document Egyptian cultural identity through the use of tools to preserve manuscripts, archives, and index materials.

Egyptian National Scientific and Technical Information Network

In 1980, the Egyptian Academy for Scientific Research and Technology embarked on a project to create a nationwide system of information services to ensure the availability and use of information and research from around the world for the socioeconomic development of the country. The result was a national network of information services in five socioeconomic sectors in Egypt, called the Egyptian National Scientific and Technical Information Network (ENSTINET). It was the first national information network in the region and has been serving the community for the past twenty years, growing with the rapid evolution in information and communications technology. One of the main functions of ENSTINET is the development and maintenance of bibliographic databases of Egyptian literature in science and technology. This database now contains 218,000 records, including the entire collection of Egyptian university theses, conference proceedings, technical reports, and journals in English, all of which can be accessed electronically. It provides the Egyptian research community with 24/7 online access to global information resources, with services ranging from online searches of databases, electronic document delivery, database development, Internet services, training, videoconferencing, and education in the field of informatics.[6]

INTERNET IN EGYPT

Internet services were first introduced in Egypt in October 1993, through a gateway established by the Egyptian Universities Network (EUN) of the Supreme Council of Egyptian Universities. Since 1994, the Egyptian domain has been divided into three main subdomains: the academic subdomain, which is served by EUN, and the commercial and governmental subdomains, which are served jointly through a partnership between the Egyptian Cabinet Information and Decision Support Center (IDSC) and the Regional Information Technology and Software Engineering Center (RITSEC).[7] Egypt's Telecom, which exercises a monopoly on basic communication services in the country, has been focusing mainly on the provision of basic communication infrastructure. However, about two years ago, it became a partner in one of the Internet Service Providers (ISPs).[8]

In 1996, the free Internet access policy was replaced by an open access policy, whereby Internet access provided to the commercial domain was privatized, and more than twelve private ISPs started operation for the first time.[9]

An ambitious project for the deployment of VSAT services for Internet connectivity was launched in 1996 to provide the rural areas with the necessary data communication infrastructure. This project complements the terrestrial solutions and helps in reducing the gap in service between well-connected regions, such as Greater Cairo, and remote and rural areas in the south of Egypt.[10]

THE REGIONAL INFORMATION TECHNOLOGY AND SOFTWARE ENGINEERING CENTER

The Regional Information Technology and Software Engineering Center (RITSEC), was established in January 1992, as a joint project between the United Nations Development Program (UNDP) and the Arab Fund for Economic and Social Development (AFESD). It is hosted by the government of Egypt—specifically, by the IDSC. It is located in Cairo. Created as a regional, nonprofit organization, RITSEC provides technical, professional, and developmental services to the agencies, institutions, and governmental organizations in the Arab Region.[11]

E-GOVERNMENT

The Egyptian E-Government Initiative aims to:

- Improve citizen services
- Create an environment conducive to investors
- Provide accurate and updated information to decision makers
- Foster Egypt's global competitiveness
- Reduce government expenditure.

The Egyptian government recognizes that one of the primary focuses of its efforts to create an Information Society must be to transform the way in which the government interacts with its citizens and with itself. For that reason, the government has made tremendous efforts to revamp its way of conducting business to begin delivering efficient, customer-focused public services. The government is committed to a fundamental reform of public services, and e-government is the primary catalyst for achieving this transformation.[12] The following are some of the Egyptian e-government sites:

- Egyptian E-Government (baouabat al-hokoma al-masria: Egyptian Government Portal) Services: [13]
  - Request a birth certificate
  - Request national ID replacement card

- Vehicular infringements
- Check phone bill
- Tourist complaints
- Exporter services
- Taxation and customs services
- Vehicle license renewal
- Electricity bills (business)
- Electicity bills (residential)
- Tanseeq (sorting students according to their marks for university admittance)
- Egypt–ALHOKOMA (Government On-Line)

  The Egyptian Cabinet Information and Decision Support Center has developed this web site with the assistance of the Egyptian Ministry of Administrative development as a first step toward e-government. The site provides information on the following sectors: government Services—Egyptian ministries—Major sites—Visitors register.[14]
- EIP: Egypt's Information Portal

  The Egyptian Cabinet Information and Decision Support Center has developed this website to provide users with following information:[15]
  - Economic Issues: government budget, banks, education, oil, external relations, industrial development, government support, population, tourism development, investments, health.
  - Databases: economic indicators, social life, education, communications and information, insurance, culture, agriculture, tourism, electricity, transportation, Suez Canal.
  - Information Infrastructure Directory—Egyptian ministries.
  - Information Publications: national data elements directory, important universal products, country information profiles, information queries, periodical pamphlets, research centers database, corporate directories.
  - Statistics: oil, agriculture, money exchange
  - Major websites
- Edara.gov: This initiative by the Ministry of Administrative Development offers online nearly seven hundred downloadable forms and procedures for different government services.[16] The site provides useful information about the ministry, administrative reform, public services, and management training.
- The State Information Service's website. The latest landmark of SIS is the creation of a website for Egypt on the Internet, joining the Super Information Highway to back up the fulfillment of its mandate with state-of-the-art language and technology. This site provides wide

coverage of current affairs as well as a broad variety of features on Egypt, with particular emphasis on Egypt's foreign and economic policies and, of course, archeological and tourist assets.

In October 2000, this site was chosen by UNESCO as one of the best cultural sites in the world. The site has so far a record of 230,814,976 visitors. As a proof of its credibility, the United States, Germany, and Japan requested that many of its sites be linked to the SIS website to benefit from its various information resources. In addition, SIS website visitors' record shows a long list of major TV and radio networks. In 2001, access to the site was launched through Nile Sat (turbo net). The site uses various multimedia tools and virtual reality technology.[17] The site provides a search engine, TV broadcast, radio broadcast, and links to the *Encyclopedia of Egyptian Arts*, an atlas of Egypt, Egyptian museums, and images and publications.

E-COMMERCE

E-commerce arrived at the end of the twentieth century compelling economists, politicians, lawyers, and bankers to rethink and reengineer work methods, policies, laws, and standards. Egypt's E-commerce Initiative can be vital in outlining important issues, raising awareness, and proposing solutions and action plans to implement such solutions. In addition, there is a lot to be gained from cooperation and exchange of experiences and expertise across borders with the rest of the world, especially with the existence of regional organizations that can support such cooperation and cross-fertilization. E-commerce as a medium for foreign trade is also a catalyst for export—implying an increase in Egypt's exports and balance of trade.

SMART VILLAGES

The Egyptian Smart Village is a center of excellence, a technology park with state-of-the-art infrastructure geared toward hi-tech businesses. The Smart Village, situated in a lush environment, offers superior Internet connection and myriad technological and administrative services, all designed to make the village an oasis for seamless IT business practices.[18]

These bodies plan for the country's future, present systems, and services, and the Egyptian Smart Village is also their creation. There are six key players planning and implementing projects to transform Egypt into an information society:

- The Cabinet—Information and Decision Support Center (IDSC)
- State Information Service (SIS)
- Ministry of Communications and Information Technology (MCIT)
- Egyptian National Scientific and Technical Information Network (ENSTINET)
- The Regional Information Technology & Software Engineering Center (RITSEC)
- The Egyptian Ministry of Administrative development.

The projects are oriented to provide Internet-based government services to Egyptian citizens; however, information literacy programs need much more intention. In addition, the coordination between the six key players is vital. Unfortunately, major Egyptian library and information science specialists as well as institutions are ignored in these projects!

TEACHING IT IN EGYPTIAN LIBRARY AND INFORMATION SCIENCE SCHOOLS

One of the crucial questions regularly put not only by IT specialists but also by common individuals and families in Egypt and other Arab countries is: What are the schools, colleges, and universities (governmental or private) that provide recognized degrees in IT applications?

The answer to that question is hard to find and often contradictory! There are nineteen academic departments teaching library and information science courses, undergraduate and postgraduate, at different Egyptian universities—for example, Cairo, Alexandria, Mansoura, Tanta, Banha, Al-azhar, Ain-shams, Helwan, Bani-souif, Almenia, Asiott, and Minofia.[19]

Cairo University—Faculty of Arts—Department of Librarianship and Documents and Information Science

Libraries and document studies in Egypt and Arab countries go back to January 1951, by Law No. 9, which establishes an institute for document and library studies at Cairo University. By the year 1954 (Law no. 611), the institute was engaged under the umbrella of Faculty of Arts at Cairo University as one of its departments. It provides undergraduate programs as well as postgraduate (master's degree, PhD, and diploma). The department has a computer lab (established in 1991) that contains twenty-five workstations connected via LAN, providing access to the

Internet. The courses provided in the undergraduate level that relate to IT applications are:

Introduction to Computers
Computer Applications in Organizing Information
Systems Analysis and Design
Information Retrieval Systems
Information Networks and Communications Technology
Database Management Systems
Programming
Publishing and Printing Technology.

The courses provided in the postgraduate level that relate to IT applications are:
Diploma and Master's Degree Level:

Computer Applications in Organizing Information
Information retrieval systems
Systems analysis and design
Communications Technology
Introduction to Computers.

PhD Level:

Electronic Information Sources
Internet in Libraries
Integrated Computer Library Systems
Computerized Cataloging
Training Technology
Communications Technology
Hypertexts and Multimedia
Electronic Publishing
Operational Research and Systems
Management Information Systems
Automated Systems for Document Centers.

The new academic legislation (2004/2005) for the department contains new undergraduate courses related to IT applications in information work, such as: Introduction to Information Technology, Electronic Data Processing, Computerized Cataloging, Electronic Publishing, Digital and Virtual libraries, Internet Usage in Libraries, and Automated Authority Control. Also, the department provides a new two-year diploma called "Automated Systems in Libraries and Information Science." The first-year courses are Data Electronic Processing, Electronic References, Database

Management Systems, Programming Languages, Systems Analysis and Design, and Indexing and Abstracting. The second year offers the following courses: Computerized Cataloging, Information Networks and Communications Technology, Integrated Computer Library Systems, Automated Authority Control Systems, Information Retrieval Systems, and Information Legislation.[20]

In addition to Cairo University's IT Department courses, it is worth mentioning two other academic departments that specialize in library and information science located at two other Egyptian universities, Minofia University and Helwan University. The first teaches the following courses at the undergraduate level: Introduction to Computers, Computers and Information Organization, Information Retrieval Systems, and Systems Analysis and Design. The second teaches the following courses at the same level: Introduction to Computers, Databases in Libraries, Automated Systems for Libraries, Information Retrieval Systems, Systems Analysis and Design, and Information Networks.

Nile University

A nonprofit academic institution that offers the best of several international universities on one campus, Nile University is beginning its first phase of operation. It is the first university in Egypt to be supported by both the government (represented by MCIT) and the private sector (represented by the Egyptian Foundation for Technology Education). The university is divided into two main schools: Nile School of Business, and the School of Engineering and Technology (NileTech). Some programs, of course, are offered in coordination between the two, such as the Management of Technology MSc program. This innovative course aims at enabling students to formulate and implement new entrepreneurial ideas and offers different areas of specialization: Communications Technology, Information Technology, Manufacturing Technology, Product Development Technology, Desert and Environmental Technology, Biotechnology, Petrochemical Technology and Heritage Information Technology.

Information Technology Institute

The Information Technology Institute (ITI) is a national institute established in 1993 by the Cabinet—Information and Decision Support Center (IDSC). It provides specialized software development programs to fresh graduates, as well as professional training programs and IT courses for the Egyptian government, ministries, and local decision support centers. In 1996, ITI launched its second branch in Alexandria to facilitate the spread of IT throughout the country, and expand the outreach of ITI.

Virtual ITI (VITI) is a leading project in Egypt and the Middle East, set for spreading IT knowledge and enhancing the training process with state-of-the-art technologies.

Aiming to bridge the gaps among researchers in the IT field, ITI organizes an annual international conference for discussing various topics in ICT. The theme of the 2004 conference (ICICT 2004) was "Multimedia Services and Underlying Network Infrastructure."[21]

Regional Information Technology Institute

The Regional Information Technology Institute (RITI) was established in 1992 as the training and professional development subsidiary of the RITSEC. Established with the aim of serving as a center of excellence in human resource development in the Arab region, the institute has continued to grow and has, throughout the years, established a sound reputation on the local and international levels for excellence in the delivery of quality training.

The activities of the Regional IT Institute are divided between two main operational divisions:

First: Academic Degree Programs (ADP)

Through a series of strategic alliances with a number of leading institutions of higher education worldwide, the Regional IT Institute offers prospective postgraduate students a variety of degree programs in areas related to business administration and information technologies. Below are the various postgraduate degree programs and their awarding institutions.

Second: Training Development Program (TDP)

TDP is the division of the Regional IT Institute responsible for the provision of training and professional development. Though the primary focus of the TDP is on quality, the figures speak highly of the department's performance throughout its relatively short history. Since its inception in 1992, the TDP has delivered more than 818 executive training programs to 10,316 participants, amounting to a massive 232,621 trainee hours.

In accomplishing these achievements, the TDP has gone beyond its mandate of serving as a regional center of excellence and has established a renowned stature in many countries spanning the four corners of the globe from the Far East to Europe and the Middle East. Overall, the institute has fulfilled the training needs of participants representing more than nine hundred organizations from over ninety countries worldwide.

Faculties of Computer and Information Sciences

By the year 1999 a new presidential decision had established faculties of computer and information sciences under the umbrella of some Egyptian universities. The faculty consists of four departments specialized in the following disciplines: Computer Sciences, Information Systems, Information Technology, and Operations Research and Decision Making.

Recently, the UNDP and the Egyptian Cabinet Information Decision Support (IDSC) and the Governorate of Sharkeya, in addition to the Investors Association–Tenth of Ramadan City have joined forces together with the United Nations Volunteers (UNV) and the Chamber of Commerce to establish three technology access community centers (TACCs) in the Governorate of Sharkeya, two hours away from Cairo. This pilot project provides rural and remote communities with public access to and a direct gateway to IT, especially the Internet, and with the training to utilize it effectively. The ultimate goal of the project is the empowerment of community members and the use of such technologies for a variety of applications benefiting the Egyptian Society. The TACCs will mainly serve as centers for the development of Sustainable Human Development (SHD)-relevant Internet content in Arabic and offer distance education programs for a variety of audiences and purposes.[22]

IT in Egyptian Library and Information Science
Research Associations and Arab Federation

Egyptian scholars in library and information science are tackling different and up-to-date research problems and publish their final works through various channels and in different forms, such as academic dissertations, conference proceedings, and journal articles. However, the majority of libraries and information centers are not applying the latest IT at work and services. Egyptian scholars are doing their best in transferring IT know-how through their published writings. They are interested in participating in seminars, meetings, and conferences in Egypt as well as other Arab countries.

In Egypt, there are four main specialized institutions that organize and supervise annual conferences. These are as follows:

1. Cairo University—Faculty of Arts—Department of Librarianship, Documents and Information Science
2. Egyptian Society for Libraries and Information[23]
3. Egyptian Society for Information Systems and Computer Technology (ESISACT)[24]
4. Internet Society of Egypt (ISE).[25]

Table 13.1. Contributions of Egyptian Scholars in the Field from 1991–1996

No.	Subject Headings	No. of Published Items
1	Artificial Intelligence	5
2	CDS/ISIS	5
3	Communications	3
4	Computer Applications in Libraries	35
5	Computer Networks	2
6	Computerized Cataloguing	3
7	Databases and Databanks	10
8	Digital Libraries	1
9	Electronic Publishing	3
10	E-Mail	1
11	Expert Systems	3
12	Hypertext	1
13	Informatics	2
14	Information Networks	8
15	Information Systems	14
16	Information Technology	15
17	Internet	14
18	Laser Discs	12
19	MINISIS	3
20	Multimedia	2
21	National Information Systems	1
22	OPACS	2
23	Selective Dissemination of Information	1
24	System Analysis	2
25	Virtual Reality Systems	2
Total No. of Items		150

Serving all Arab nations are two main institutions:

1. The Arab Federation for Libraries and Information (AFLI), which holds annual conferences as well as workshops to enhance the skills of personnel working in libraries and information services[26]
2. Special Libraries Association, Arabian Gulf Chapter.[27]

IT Applications in Information Work Research Trends

What kind of IT applications in information work are manipulated, discussed, and published by Egyptian scholars? To find an answer to this question, the most famous Arabic bibliographic tool that aims at regular coverage of Arabic literature in library and information science was inspected.[28]

Table 13.2.   Contributions of Egyptian Scholars in the Field from 1997–2000

No.	Subject Headings	No. of Published Items
1	Communications	17
2	Computer Applications in Libraries	37
14	Computerized Cataloguing	5
11	Computers	4
16	Databases and Databanks	14
17	E-Books	3
8	E-Commerce	2
12	E-Journals	4
19	Electronic Information Sources	7
15	Electronic Sources Cataloguing	3
7	E-Mail	2
21	E-Publishing	8
23	Expert Systems	3
22	Hypertext	3
20	Informatics	3
10	Information Technology	28
5	Internet	57
4	Laser Discs	8
13	Libraries and Information Networks	18
18	Metadata	1
3	MINISIS	2
25	Multimedia and Hypermedia	3
6	Software	4
9	Systems Analysis	2
24	Virtual Reality Systems	1
Total No. of Items		239

After excluding the items written by researchers from other Arab countries, as well as items handling traditional or regular library and information science topics, the total contribution of Egyptian scholars in the field during the years 1991–1996 will be as shown in table 13.1.

The subjects that attracted the majority of scholars during that period (1991–1996) are ranked as follows:

1. Computer applications in libraries
2. Information technology
3. Information systems
4. Internet
5. Laser discs.

After excluding the items written by researchers from other Arab countries, as well as items handling traditional or regular library and information science topics, the total contribution of Egyptian scholars in the field during the years 1997–2000 is as shown in table 13.2.

The subjects that attracted the majority of scholars during that period (1997–2000) are ranked as follows:

1. Internet
2. Computer applications in libraries
3. Information technology
4. Libraries and information networks
5. Communications
6. Databases and databanks.

Finally, if we compare the results of the two periods we may end with the following conclusions:

- The total number of published items has increased remarkably. There were 150 items published during the years 1991–1996, but the total number reached 239 during the years 1997–2000.
- Major topics and issues handled by scholars are Internet, Computer Applications in Libraries, and Information Technology.
- New IT subject headings added to the bibliography during the period 1997–2000 are as follows: E-Books, E-Commerce, E-Journals, Electronic Information Sources, Electronic Sources Cataloging, and Metadata.

IT IN EGYPTIAN LIBRARY AND INFORMATION SCIENCE PROFESSIONAL WORK

Integrated Library Systems

Library automation in Egypt goes back to the year 1961, when the documentation center related to the National Planning Institute established with the assistance of a computer company, a bibliographic relational database using FoxPro software. Later, different library automation projects were implemented in Alexandria University Libraries, the Egyptian national library, and others.[29, 30]

The twenty-first century has witnessed different automation projects in Egyptian libraries based on two types of systems. The first type consists of Egyptian-originated systems such as the a-LIS system. The second type, Arabized foreign-originated systems, include such as Horizon (AL-Ofuq), VTLS (Virtua), UNICORN, and INNOVATIVE.

A-LIS, Advanced—Library Information System, created and distributed by IDSC, first appeared in 1989 as LIS-1; then in 1991 a new version,

LIS-2, appeared, followed by LIS-3 in 1996. By May 1999 twenty-six Egyptian libraries had implemented the system and forty-two libraries had requested it.[31]

Some of its main features are:

- Web access, enabling database searching, e-mail, current awareness, reservations, materials borrowing
- Bar code facilities
- Multi database management system support
- Fully security control, including user rights and functions
- In compliance with software quality control rules
- Version control rules applied
- Full Arabic support
- Trilingual interface
- Labels printing utility
- Full integration of all library transactions
- Multi choice compound search facility
- Import and export data with similar databases
- Automatic claiming facility
- Unification of authority lists data
- General reports of daily library transactions.

The operating platform is Windows 9x/NT. Currently 458 libraries are using the system, among them the libraries of the Ministry of Planning, the IDSC, Asuit University, Youth Centers, and the Integrated Care Society, as well as Greater Cairo Public Library, Hileopolis Public Library, and some school libraries.[32]

Horizon (AL-Ofuq) Integrated Library System, Arabized and distributed by Arabian Advanced Systems for most of the Arab countries, including the United Arab Emirates, Saudi Arabia, Qatar, Oman, Kuwait, Oman, and Egypt. The system supports Public Access Catalogue, Cataloguing, Circulation, Serials Control, Acquisition, and System Administration.[33, 34]

In Egypt, the system is working at Al-Zagaziq University, Banha University, and Al-Mansoura University.

VTLS (Virtua) features its newly introduced Library Automation in 3V. This three-part offering is composed of Virtua ILS—Integrated Library Systems, Visual MIS—Multimedia & Imaging Solutions, and Vista CPS—Companion Product Suite.[35] Virtua ILS—Integrated Library Systems includes the company's third-generation library automation system, which has recently been enhanced with a Web-based reports management system.

With Visual MIS—Multimedia and Imaging Solutions, VTLS has integrated its digital library offerings, collection management, and data-entry services. Its Vista CPS—Companion Product Suite encompasses many new and exciting products and services that have resulted through partnerships with other visionary companies in the industry in the development of VTLS's Radio Frequency Identification (RFID) technology.

In Egypt, the system was working at the Egyptian National Agriculture Library before its decision to transfer to UNICORN system.

UNICORN, Sirsi's Unicorn Library Management System, provides librarians with a total management infrastructure for directing all aspects of their libraries. Designed to enable libraries to meet changing technology head-on, Unicorn is the library system for today and into the future, featuring:

- A flexible multitiered client/server architecture for libraries of all types and sizes.
- Open architecture that permits the library to incorporate new technologies into the system.
- The ability to maximize the power of the "e-Library Electronic Library," which enables libraries to meet the demands of today's information seekers.
- WorkFlows staff client, with next-step technology to guide staff through various tasks and related activities.
- Powerful add-on modules that support all of the library's public and technical services.
- Complemented by Sirsi's Innovative Hyperion™ Digital Media Archive for managing digital collections. Mubarak Public library was using ALEPH (Automated Library Expandable Program) before transferring to UNICORN.[36]

Innovative Interfaces partners with libraries worldwide to provide Web-based information technology solutions to both patrons and staff. The Innovative Millennium system is a Web-based, open-platform integrated library system that offers the best and most comprehensive functionality of any library automation software. Its Java technology–based interface offers staff and patrons an intuitive, easy-to-use, and platform-independent system. With its multitiered system architecture, object-oriented design, and complete scalability, Millennium provides full, integrated functionality; its core modules constitute a time- and library-tested automation system that can be implemented in every type of library.

In Egypt, the system is working at the American University in Cairo Library.[37]

In addition to the mentioned systems, libraries can choose ACLIB from Automation Company and CDS/ISIS from UNESCO.

Digital Library Projects

There are various digital library projects planned or implemented targeting digitization of printed or AV materials. Mubarak Public Library's digital library provides some electronic full-text books related to computer science and Internet.[38] On October 16, 2002, the new Bibliotheca Alexandrina was inaugurated. After years of effort, the library complex is now open to the public. UNESCO has played an instrumental role in the revival of the library, with its foundation stone being laid by its director general and the president of Egypt in 1988. The largest library in the Arab world, its collection capacity is eight million volumes.[39, 40] The Digital Modern History of Egypt was started by the Bibliotheca Alexandrina to integrate digital libraries about Egypt's modern history. It will include the collection of specialized libraries, such as those of Gamal Abdel Nasser, Anwar El Sadat, and eminent Egyptian authors and historians, as well as contents related to the subject from all over the world. The objectives are to scan, catalog, index, and OCR the collection and present it in a searchable form to its users. This project will benefit from the efforts made for the Million Book Project.

There is an encyclopedic effort to use ICTs for archiving and accessing rare treasures of Arab and Islamic civilization, called "Contributions of the Arab/Islamic Civilization to the Sciences." It is being developed in Cairo through a joint cooperative project of the UNESCO Cairo Office, the Egyptian National Center for Documentation of Cultural and National Heritage (CULTNAT), and the National Library of Egypt. The first title, "Arabic/Islamic Contributions in the Field of Medical Sciences," has been published. The main result of this first phase of the project was the digitization of two thousand manuscripts that are available in the public domain for the benefit of researchers, analysts, and other interested readers. The totality of the digitized collection will be published on a trilingual (Arabic, English, French) CD-ROM as well as on the Internet.[41]

Egyptian Library Websites

Although most Egyptian institutions are keen to have a website and are looking for a good position in cyberspace, Egyptian library websites lack major electronic features like online interactivity, integrated digital services, and regular updating. The American University in Cairo (AUC) library website is a good model for university library websites; it provides webPAC, online database searching, Ask-a-Librarian , and other services.[42] Mubarak Public Library is a good model for public library websites, providing webPAC, online databases searching, and current awareness updating services.[43]

Egyptian Professional Websites

Egyptian Libraries Network (ELNet) and Directory

The IDSC Library introduces ELNet, which contains online databases of the automated libraries in Egypt.[44] A user can search a single library database, or use the union catalogue to search all online Egyptian library databases. The user can not only search the libraries catalogue but also search for any information about these libraries through the directory of the Egyptian libraries.[45]

Cybrarian—Arabic Portal for Libraries and Information

Cybrarian is a portal that supports Arab librarians and academics as well as scholars with valuable, up-to-date information related to library websites, Arab publishers, associations, and organizations, e-journals, library schools, automated library systems, reference sources and services, and Arab professional conferences, seminars, and meetings. In addition to the mentioned services, the portal has started publishing a free-refereed professional e-journal.[46]

Librarian Net—Information Specialists and Librarians Network

Librarian Net exists mainly to provide e-journals specializing in librarianship and information science. In addition it provides current information about conferences, news, and different issues.[47]

THE IT PRIVATE SECTOR IN EGYPT

The Egyptian private sector participation in the communication sector is presented in table 13.3.[48]

Table 13.3. The Egyptian Private Sector Participation in the Communication Sector

Service Type	October 1999	May 2004
Internet Service Providers	40	176–262 number
Mobile	2	2
Public Data Network	1	7
Internet Backbone Providers	1	4
Value-Added Voice Services	—	2
Public Phone Services	3	2
Equipment Manufacturing	2	3
Copper Wire Manufacturing	3	4
Fiber Optic Cables	2	2
VSAT	—	2
GMPCS	—	2

CONCLUSION: MISSING PIECES

- Egypt does not yet have a government communications network whereby different government bodies can interact and exchange information and documents.
- Egypt lacks standards and specifications for the process of government automation.
- Egypt's high rate of information illiteracy is a challenge hindering the majority of the population from benefiting from the new model of e-government.
- Success in the library profession in practice (specially in the Arab countries and Egypt in particular) depends on close attention to specialized research topics and trends as well as to library and information science education programs, as shown in the following triangle:

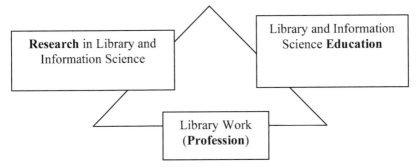

Figure 13.1.   How to Achieve Success in the Library Profession

The field of library and information science has changed during the last thirty years. In the twenty-first century, librarians must get used to the changing environment and learn some basic essentials in the field including the following:

- Personal characteristics such as versatility, social skills, and professional and academic distinction, in addition to managerial ability, are prerequisites for library and information work.
- Any library or other information utility should take into account that, although its physical location has to be in one country, its electronic existence (virtual location) through its website is global. That means there are universal standards, protocols, legislation, and systems specifications that should be respected, adopted, customized, and tailored to our needs.
- No single library looking to satisfy users' needs can work independently! Nor do they have to; cooperative projects in various library activities have been facilitated through communications technology.

- Internet and ICT will play crucial roles in helping libraries and information utilities to achieve their aims.

NOTES

1. The Egyptian Information Society Initiative/ Ministry of Communications & Information Technology. www.mcit.gov.eg (accessed June 4, 2004).

2. The Egyptian Information Society Initiative.

3. The Egyptian Cabinet Information and Decision Support Center (IDSC). www.idsc.gov.eg (accessed June 4, 2004).

4. Welcome to SIS—Egypt State Information Service. www.sis.gov.eg (accessed June 4, 2004).

5. Ministry of Communications and Information Technology. www.mcit.gov .eg/ (accessed June 8, 2004).

6. Egyptian National Scientific and Technical Information Network (ENSTINET). www.sti.sci.eg/ (accessed June 9, 2004).

7. Regional Information Technology and Software Engineering Center (RITSEC). www.ritsec.org.eg/ (accessed June 9, 2004).

8. Sherif R. Hashem, *Technology Access Community Centers in Egypt: A Mission for Community Empowerment* (Cairo: The Regional Information Technology & Software Engineering Center, 1999).

9. Sherif Hashem and Magda Ismail, *The Evolution of Internet Services in Egypt: Towards Empowering Electronic Commerce* (Cairo: The Regional Information Technology & Software Engineering Center, 1998).

10. S. Hashem and T. Kamel, *Paving the Road for Egypt's Information Highway: Proceedings of the First Kuwait Conference on the Information Highway* (Kuwait: Institute for Scientific Research, 1998).

11. The Regional Information Technology & Software Engineering Center www.ritsec.org.eg/html/profile.html (accessed June 15, 2004).

12. Ministry of Communication and Information Technology. EISI-Government The Egyptian Information Society Initiative for Government Services Delivery. (7/3/04AD-1/11). www.mcit.gov.eg/ (accessed June 16, 2004).

13. Egyptian e-government. www.egypt.gov.eg/english/services/ (accessed June 10, 2004).

14. Egypt—ALHOKOMA(Government On-Line). www.alhokoma.gov.eg/ (accessed June 10, 2004).

15. EIP: Egypt's Information Portal. www.eip.gov.eg/ (accessed June 12, 2004).

16. Edara.gov. www.edara.gov.eg/ (accessed June 12, 2004).

17. Welcome to SIS—Egypt State Information Service. www.sis.gov.eg (June 18, 2004).

18. The Egyptian Smart Village. www.smart-villages.com/whatisit.htm (accessed June 18, 2004).

19. Department of Library and Documents and Information science, Faculty of Arts, Cairo University, "Department of Librarianship and Documents and Information Science through 50 Years (1951–2000): Memorial Book for the Golden Day of the Department" (Cairo: Academic Library Publisher 2002). In Arabic.

20. "New Academic Legislation for Cairo University Dept.—*Cybrarians.*" www.cybrarians.info/main.htm (accessed June 20, 2004).

21. ITI: Information Technology Institute. www.icict.gov.eg (accessed June 20, 2004).

22. Hans D'Orville, *Towards the Global Knowledge and Information Society, the Challenges for Development Cooperation—INFO21* (IT for Development Programme, 2004).

23. Egyptian Society for Libraries and Information. www.mans.edu.eg/libr/ELA/index.htm (accessed June 18, 2004).

24. Egyptian Society for Information Systems and Computer Technology (ESISACT). www.esisact.com.eg/main.html (accessed June 19, 2004).

25. Internet Society of Egypt (ISE). www.ise.org.eg/ (accessed June 19, 2004).

26. The Arab Federation for Libraries and Information (AFLI) (accessed June 19, 2004).

27. Special Libraries Association, Arabian Gulf Chapter. www.agu.edu.bh/sla-gc/ (accessed June 21, 2004).

28. The most famous Arabic bibliographic tool that aims at regular coverage of Arabic literature in library and information science, published through the following integrated works:

- Mohammed Fathi Abdel-hadi. Library and Information Science Arabic Literature (Riyadh: Dar-Al-marich, 1981).
- Mohammed Fathi Abdel-hadi. Library and Information Science Arabic Literature in Ten Years: 1976–1985. (Riyadh: Dar-Al-marich, 1989).
- Mohammed Fathi Abdel-hadi. Library and Information Science Arabic Literature 1986–1990 (Riyadh: King Fahad National Library, 1995).
- Mohammed Fathi Abdel-hadi. Library and Information Science Arabic Literature 1991–1996 (Riyadh: King Fahad National Library, 2000).
- Mohammed Fathi Abdel-hadi. Library and information Science Arabic Literature 1997–2000 (Riyadh: King Fahad National Library, 2003).

29. Amal wagih Hamdy, "Automated Systems Installed in Arabic Libraries and Information Centers (part-1)," *Arab Journal of Library and Information Science* 20, no. 3 (July 2000): 143–69.

30. Amal wagih Hamdy, "Automated Systems Installed in Arabic Libraries and Information Centers (part-2)," *Arab Journal of Library and Information Science* 20, no. 4 (October 2000): 118–45.

31. Usama Al-said Mahmoud, "Criteria for Selecting and Evaluating Integrated Computer Library Systems: An Applied Study on the Latest Version of the A-LIS/," *New Trends in Libraries and Information Science* 7, no. 13 (January 2000): 129–68.

32. "A-LIS: Advance Library Information System: English-French-Arabic," www.alis.idsc.gov.eg/ (accessed June 10, 2004).

33. "Arabian Advanced Systems." www.aas.com.sa (accessed July 4, 2004).

34. Abdl-Gabar Al-Abdl-Gabar, Mohammed Moaoud. "AL-Ofuq (Horizon): Automated System for Administering Arab Libraries," *Arabic Studies in Libraries and Information Science*, no. 3 (July 1996): 188–209.

35. VTLS, Inc. www.vtls.com (accessed July 2, 2004).

36. Hane Muhi Al-deen Atieh, "Integrated Computer Library Systems in Arab Libraries between the Theme and Technology: Introduction to ALEPH System," *Arabic Studies in Libraries and Information Science* 5, no. 2 (May 2000): 13–18.

37. "Information Technology Advances in Libraries /Sun Microsystems, Inc. 2003." www.iii.com (accessed July 5, 2004).

38. "Welcome at Mubarak Public Library–Digital Library." www.mpl.org.eg/english/dig-lib.htm (accessed August 5, 2004).

39. "Libraries: Bibliotheca Alexandrina Reopens."—*UNISIST Newsletter* 30, no. 2 (February 2002): 5.

40. "Bibliotheca Alexandrina: Front Page." www.bibalex.gov.eg/English/index.aspx (accessed August 4, 2004).

41. "Arab States: Digitization of Islamic Scientific Manuscripts," *UNISIST Newsletter* 30, no. 2 (February 2002): 19.

42. "American University in Cairo (AUC) Libraries." http://libhelp.aucegypt.edu/libweb2/index.htm (accessed August 4, 2004).

43. "Mubarak Public Library." www.mpl.org.eg (accessed August 5, 2004).

44. "Egyptian Libraries Network." www.library.idsc.gov.eg/ (accessed August 1, 2004).

45. "Directory of the Egyptian Libraries." www.libdirectory.idsc.gov.eg/main.html (accessed August 2, 2004).

46. "Cybrarian—Arabic Portal for Libraries and Information." www.cybrarians.info/main.htm (accessed August 6, 2004).

47. "Librarian Net Dot Com." www.librariannet.com/ (accessed August 6, 2004).

48. Ministry of Communication and Information Technology. www.mcit.gov.eg/ (accessed August 6, 2004).

Chapter 14

The Impact of Technology on Middle Eastern Collections and Services in the United States

Ali Houissa

ABSTRACT

High-speed communication and computer processing capabilities are revolutionalizing information gathering and dissemination in the field of Middle Eastern studies, impacting the ways in which data are acquired, recorded, inventoried, transmitted, and published. This chapter describes, in addition to a brief history of Middle East collections in the United States, some of the latest technical innovations and tools that libraries with Middle Eastern collections use, with particular emphasis on the Wide World Web and the digital environment, metadata, Unicode and the vernacular in integrated library systems, and cooperative resource sharing.

INTRODUCTION

As digital technology leaps forward, high-speed communication and computer processing capabilities are increasingly impacting information gathering and research in the humanities and social sciences. Libraries are in the forefront when it comes to harnessing these developments to support a diversified client base and to improve the user's access to research tools such as full-text documents, bibliographies, indexes, and audiovisuals. The Internet, by collapsing the geographic divide, makes it more convenient and affordable to access a growing wealth of information.

219

Area studies in general have for decades been inconvenienced by geography, and that has been particularly true of the study of the Middle East. The ways in which data about the region are acquired, recorded, inventoried, transmitted, and published have been witnessing enormous change over the last several years. American libraries and other institutions are pioneering efforts to provide nontraditional means of access to information, even in the vernacular scripts. In this chapter I consider some of the ways in which technology has been introduced in the context of Middle Eastern library collections and services in the United States. Without dwelling on technical minutiae or the plethora of computer applications, products, and multimedia available, I will describe some of the latest standard methods that libraries with Middle Eastern collections use, with particular emphasis on the following: the Wide World Web and the digital environment, metadata, and cooperative resource sharing.

LIBRARY COLLECTIONS AND SERVICES FOR THE MIDDLE EAST IN THE UNITED STATES

There are several distinguished Middle East collections in the United States, some of which are regarded as among the best in the world and as major resource centers for the study of the Middle East, at least from the point of view of the breadth and comprehensiveness of their coverage. Varying in age widely, many of these collections—in university, large public, or special and private libraries—have custody of substantial material in the non-Roman-alphabet languages of the region that includes North Africa, the Arab world, Turkey, Iran, and Afghanistan.[1]

For a long time, little attention was paid to study of the Middle East in this country, except by anthropologists, archaeologists, and probably a few others; the region was seen primarily through the prism of Cold War competition. Major post–World War II events, such as the Arab-Israeli conflicts, the Iranian Revolution, and oil crises, made this pivotal area a focal point for studies and research—both historical and contemporary. In the second half of the twentieth century, Middle East library collections underwent an expansion similar to that in the rest of the scholarly universe, as government funds were directed to area studies, among them those promoting programs on the Middle East. Several "study centers" and the research libraries associated with them began to appear at universities with funding through renewable grants for book acquisitions and support for staff salaries. Since 2002 most of these centers have received a boost in federal funding. There has also been a major transformation of Middle East studies programs at many colleges and universities in the last few years, revitalized through increased enrollment, faculty hiring, diversified course offerings, and large endowments. An attendant transformation of library collections and services has been taking place: holdings already significant in such fields as history, religion, literature, economics, linguistics, and the arts are being improved, and access to them is being reshaped with the ever-increasing number of source texts, journals, newspapers, and indexes that are of interest to researchers becoming available in electronic format. As in other fields, along with the use of the Internet in classrooms, library online databases and other electronic resources are being increasingly integrated into teaching; mastery of automated library research, not to mention acquisition of the necessary hardware and connectivity, are nowadays prerequisites that every student or faculty member is expected to possess.

AUTOMATION, DIGITALIZATION, AND THE VERNACULAR

Rapid and in-depth access to information usually justifies investing in information technologies in the library world, and it is the "digital library" that is the most noteworthy manifestation of that investment. Institutions collecting in major Middle Eastern languages—mainly Arabic, Persian, and Ottoman Turkish[2] (Hebrew, Ancient and Modern, for significant Hebraica collections)—have for some decades wrestled with a serious challenge: how to represent a nonroman script, in the online catalog for instance, short of reproducing it in the vernacular. Established romanization systems work fine for languages that are rendered letter for letter.[3] It is, however, a far less simple matter for languages with nonvocalized alphabets that are usually determined by the reader from context. Moreover, even with the meticulous and proper romanization from the Arabic or the Hebrew alphabet (nonvocalized), isn't that in reality trying to tailor one thing to fit into something else of a completely different shape and form, running the risk along the way of denaturing the original? That is one of many reasons why library catalogs resort to heavy cross-referencing to account for all the possible manifestations of a word and for guidance, that is, transliterating a single word in different ways. For any system to become fully meaningful, it must, therefore, support the vernacular, or at least allow for nonroman characters interfacing. The fact that languages using the Arabic or Hebrew alphabets are written right to left does not, of course, alleviate complexity—more likely it would accentuate it.

Transliteration of the vernacular into Latin script was for a while the only possible way to encode machine-readable bibliographic records, as ASCII was the only available electronic character set. In the late 1980s and early 1990s the USMARC character sets for Arabic[4] were introduced and implemented in bibliographic records.[5] Major bibliographic utilities such as the Research Libraries Group (RLG) and the Online Computer Library Center (OCLC) added Arabic to their systems. RLG's Research Libraries Information Network (RLIN) started including vernacular fields in 1991 with search and display in Arabic script.[6] In 2004 RLG released the RLIN21 Web interface for searching its union catalog and authority files, also allowing retrieval and export of MARC records. It displays Arabic-script records when used with a Unicode-supported browser. There's also the RLIN21 Windows client with Arabic-script support, released later the same year, which can be used to create and update bibliographic and authority records.

OCLC has also developed its own Arabic cataloging software, a free, Windows-based application that allows cataloging in Arabic script. It makes it feasible to enhance existing romanized records with Arabic characters by using a special transliteration tool that automatically converts the roman data to Arabic. As of April 2004, OCLC's database, WorldCat, has almost 300,000 "Arabic-language-coded" records (LANG = ARA) with or without vernacular script and fifty-seven thousand records[7] that contain Arabic script characters. Several large libraries in the Middle East have agreed to contribute their Arabic-language holdings information to WorldCat. In the summer of 2004 OCLC expanded its capabilities to offer full Arabic cataloging (outsourcing), both vernacular and romanized. This has been accomplished with the formation of an Arabic cataloging unit overseas—in the Middle East. OCLC also partners with vendors to include pre-catalog materials in Arabic. Users of RLIN or OCLC are required to follow their respective guidelines and manuals for cataloging Arabic materials.

Unicode and Integrated Library Systems

Unicode is a universal standard for representing multilingual texts and different writing schemes of the world's major languages to allow text data to be interchanged, sorted, searched, and manipulated consistently and without conflict. In the library environment, one would of course need a platform, such as an online catalog system, that can manipulate the character set and the bidirectional algorithms in the case of languages that read from right to left such as Arabic and Hebrew. There are now integrated library systems with modules supporting Unicode for various languages, developed by a number companies, offering a range of features for indexing, cataloging, searching, and retrieval and including library management features for circulation, reserves, interlibrary services, and acquisitions.[8]

VTLS, Inc. (Visionary Technology in Library Solution) provides the VIRTUA integrated Library Management System, a comprehensive client-server software for library collections that allows Web access to the library's online catalog and fully complies with the Unicode standard on both the client and server levels. The system stores data natively in Unicode, making it possible to import, catalog, and display records in any language and also to change interface languages on the fly (i.e., in real time) and store multiple scripts within the same MARC tag or subfield.

A Web-based library automation system called "III" by Innovative Interfaces, Inc., has modules for various languages and is able to handle a variety of foreign scripts for both display and search, including Hebrew and Arabic scripts. The system basically receives the characters in Unicode and then uses a table to translate them into an III-based mapping. For users to search and display Arabic or other scripts, they must have the appropriate fonts loaded in their computer. Thus, Arabic script records that include the vernacular fields can be downloaded from OCLC through an online interface to a local system in real time without rekeying.

Endeavor Information Systems, an associate member of the Unicode Consortium, introduced Unicode capabilities in the 2000.1 release of its Voyager system, which consisted of dynamic conversion of data in the MARC 880 field to the Unicode character set for display in WebVoyáge, the Voyager OPAC. Endeavor's ENCompass 3.5 system for managing, searching, and linking collections is currently being tested. It has Unicode support and can handle non-roman scripts (including Arabic, Hebrew, Persian, etc.) and convert bibliographic, holdings, and authority records in Voyager to the Unicode UTF8 standard.[9]

The Internet, Metadata, and Digital Libraries

With the implementation of the cross-platform Unicode standard, it is now also possible to do word processing, desktop publishing, Internet browsing, and e-mail and to access full-text documents in online databases or on CD-ROM in Middle Eastern languages. That's because texts saved in this encoding will open, in general, without conversion on individual applications (Windows, Macintosh, and Unix). Microsoft's Windows 2000 and XP operating systems and Microsoft Office (2002, 2003, .NET) are Unicode-enabled with support for right to left languages (Arabic, Hebrew, Persian) and the ability to combine them with "European" fonts in a single document.[10]

The Macintosh O/S X system also comes with various utilities and is fully functional for reading and searching in Arabic and many other languages. System Macintosh O/S X 10.2 is code-named "Jaguar," and the newer O/S X 10.3 is code-named "Panther." Next in line to be released is OS X "Tiger").

Linux is another international operating system, with utilities and device drivers that support multilingual character sets, including non-Latin alphabets such as Arabic and Hebrew. (Macintosh O/S X systems have a UNIX-based foundation.)

In general, all the Web standards (HTML, XML, etc.) support or require Unicode. The latest versions of Netscape Navigator and Internet Explorer both support Unicode.[11] The World Wide Web Consortium (W3C) has created the HTML and CSS (Cascading Style Sheets for xHTML documents) standards with items specifically to correctly display Middle Eastern languages. Many Web browsers now have support for Middle East content, either natively or through the use of plug-ins.

What hypermedia, hypertext, and multimedia can do now is digitally merge very large amounts of unassociated materials of different kinds and formats (text, data, graphics, pictures, sounds, video, applications, etc.) that then can be accessed on demand across the Internet channels or distributed on CD-ROM and other portable storage devices. These and further developments are significantly changing research, scholarly communication, the exchange of ideas, and publishing in the field of Middle East studies—as well as in other humanistic and social sciences fields. Metadata, or information about information, is a way of describing and summarizing information about various aspects of electronic documents such as content, quality, format, field of knowledge, copyright issues, etc. As structured information, metadata can improve search engine results by widening the scope for locating useful information, usually through the use of a more controlled vocabulary and by providing a precise context for the words. Bibliographic metadata is rich and multilingual, with the ability to handle multiple (non-Latin) scripts. Encoding of bibliographic metadata used to be limited by the computers' inability to handle various vernacular character sets, until a system was developed that could handle records in non-Latin scripts (Japanese, Arabic, Chinese, Korean, Persian, Hebrew, and Yiddish, or "JACKPHY"). It is only Unicode's universal character set that is truly enabling bibliographic metadata to integrate the digital world. Scripts written right to left can not only display across properly equipped platforms, they can do so with any number of other scripts (multilingual sources).[12]

The abundance of Middle East-oriented websites[13] and other Internet-based venues, such as newsgroups, archives, mailing lists, etc., are widening the spectrum of available information. However, as more information is available than ever before on the virtual "wasteland" that the Internet can be, there's a greater burden on librarians and information specialists to sort out and vet what is of value.

The Web presence of Middle East Studies programs and research centers in American institutions is now considered a vital, integral part of their existence. For library collections with Middle East holdings, computers and the Internet have made the most impact in areas that require rapid but accurate storage and processing of information and the ability to connect, communicate, and operate twenty-four hours a day.

There now exist growing numbers of online gateways to sources on individual countries and on Islam and Islamic history and culture—handbooks, encyclopedias, and so on. One could reasonably imagine that few users would prefer going back to bulky, multiple-volume printed versions of dictionaries, encyclopedias, indexes, or full-text sources. Think, for instance, how expedient it is to browse the monumental works in the field: *Index Islamicus* and the *Encyclopaedia of Islam Online*. Both are valuable multidisciplinary sources now accessible in digital format, the *Encyclopaedia* with well over thirteen thousand journal articles and the *Index* by far the most important bibliographic database of books and articles about Islam and the Islamic world.

An online project of interest was launched in 2002 by the American Council of Learned Societies. In conjunction with ten university presses and eight learned societies, including the Middle East Studies Association, the History E-Book (HEB) Project[14] involves publishing historical monographs online. These are "works of major importance to historical studies" that are frequently cited in the literature. Currently, there are nearly eight hundred titles listed, but the project plans to add approximately 250 books annually to the collection, "as well as the balance of eighty-five new electronic titles that have the potential to use Web-based technologies to communicate the results of scholarship in new ways." These e-books are accessible through subscribing libraries and their learned societies. Moreover, this pioneering project offers an opportunity for scholars to plan and write e-books and for scholarly presses to develop expertise by experimenting with electronic publishing. For libraries it means worthy additions to their emerging quality e-collections—possibly at reduced cost.

MENALIB,[15] a virtual library for the Middle East and North Africa, is another information portal for Middle Eastern and Islamic studies, with links to over 1,600 Internet sites and its own search engine (al-Misbah). Started in October 2000 at Halle University Library and funded by the German Research Foundation (DFG), it covers the entire Middle East, including Afghanistan, the Caucasus, Central Asia, and the Horn of Africa but excluding Israel, except for resources dealing with the Arab-Israeli conflict in general.

Library Consortia

An ambitious collaborative database project was jump-started in late 2002, named Online Access to Consolidated Information on Serials

(OACIS) and led by the Yale University Library with the participation of eight other universities, to improve access to Middle Eastern serials in libraries in the United States, Europe, and the Middle East (initially for Arabic and English-language titles and then for an expanding group of Middle Eastern languages). It is funded for three years through a U.S. Department of Education Title VI grant under "Technological Innovation and Cooperation for Foreign Information Access." The ultimate purpose is to create a publicly and freely accessible listing of Middle East journals and serials, including periodicals in print, microform, and online, and to serve as a gateway by enhancing content delivery of titles. The listing identifies libraries that own the materials as well as exact holdings. The database is now available on the Web.[16]

The Center for Research Libraries (CRL) is an older consortium of North American universities, colleges, and independent research libraries that acquires and preserves traditional and digital resources for research and teaching and makes them available to member institutions through interlibrary loan and electronic delivery. Its emphasis is on collections of "low use or less commonly held materials." CRL has collected and preserved Middle East studies research materials for more than fifty years, mainly as part of its area studies microform projects. The Middle Eastern Microform Project (MEMP), established in 1987, acquires microform copies of unique and research material in the field of Middle Eastern studies and preserves rare prints and manuscripts. The coverage includes the Arab Middle East, Israel, Turkey, Iran, Afghanistan, Central Asia, and "other related areas not covered in other cooperative microform projects."[17]

The center has also launched the Middle East Research Journals project (MERJ) in collaboration with the Council of American Overseas Research Centers (CAORC) to provide bibliographic access, preservation microfilming, article-level indexing, and digital document delivery for more than 2,200 Middle Eastern journal titles. Many of these titles have been converted into MARC format and made available through the American Overseas Digital Library Online Union Catalog.[18] It was initiated by CRL as part of the Mellon Foundation–funded investigation of archiving political communications from the World Wide Web, an examination of the scholarly uses of Web content as primary source materials for research and publication such as Web-based political communications related to the Middle East.[19]

The National Institute for Technology and Liberal Education (NITLE) initiated in 2001 a collaborative Internet project called Arab Culture and Civilization,[20] an ongoing effort focusing on the Islamic world as a whole and regionally on the twenty-two Arab countries of the Middle East and North Africa. The material includes original texts, video clips, and audio files from online and print sources, organized in ten different categories or "modules" that are cross-linked: History; Ethnicity and Identity; Islam; Arab Americans; Literature and Philosophy; Popular Culture and the Performing Arts; Family and Society; Art and Architecture; The Arabic Language, Geography, Demographics, and Resources; New Media. It's principally geared toward the English-speaking liberal arts colleges served by NITLE, but the site is open to all visitors, and it is technically structured to be accessible across the widest possible range of browsers and operating systems.

CONCLUSION

While the promise of rapid and in-depth access to information is not yet fulfilled in most Middle Eastern countries, as pointed out in *The Arab Human Development Report, 2002*,[21] many of these countries, as well as other developing regions, look up to North American institutions as pioneers in digital innovations and applications. Many U.S. libraries with significant area studies holdings are engaging in serious cooperation with an increasing number of Middle Eastern libraries and institutions, as illustrated by the example of the OACIS project. The need for investment in technology is fueled by a rapidly wired world. Libraries in general, and those among them with special vernacular collections in particular, are painfully aware that they do not yet possess all the answers with regard to automation, digitization, or making their holdings accessible via the Internet and on the Web. Multilingual information is certainly becoming easier to access thanks to universal character sets like Unicode. However, despite the rapid pace of technological change, many aspects of its application remain to be grasped and applied in the library environment. For instance, what is the best way to offer meaningful computer-based learning platforms, e-journals and e-books, pay-per-view Web services, and digital information that can be seamlessly charged back to the library?

NOTES

1. There is lack of consensus on one single definition of the "Middle East" or "Near East" region.

2. All three are written in the Arabic alphabet. However, these languages are not related linguistically. Arabic is Semitic, Persian is Indo-European, and Turkish is Ural-Altaic (Turkic).

3. Such as languages written in the Greek, Cyrillic, or Armenian alphabets. Vocalization is usually not supplied or obvious in the Arabic alphabet, so the reader must supply the vowels, which requires a degree of practical knowledge of the

languages. See the 1997 edition of the *ALA-LC Romanization Tables: Transliteration Schemes for Non-Roman Scripts.*

4. Basic and an extended version of the character sets are meant to support conversion of other vernacular languages using the Arabic script: Kashmir, Kurdish, Ottoman Turkish, Pushto, Sindhi, Uighur, and Urdu.

5. Over the years, the Machine-Readable Bibliographic Information committee (MARBI), [www.ala.org], has routinely developed and improved USMARC standards for non-roman scripts in bibliographic records. The basic USMARC Arabic script set was based on two standards, ASMO Standard Specification 449 (of the Arab Organization for Standardization and Metrology), and ISO 9036 (Information Processing—Arabic 7-bit Coded Character Set for Information Interchange). The extended USMARC Arabic script set was developed at the same time as ISO 11822 (Information and Documentation—Extension of the Arabic Alphabet Coded Character Set for Bibliographic Information Interchange). The ISO extended Arabic script set is completely synchronized with USMARC.

6. Book titles by language and publication date (RLIN—March 2004):

Arabic:		Persian:	
All	With	All	With
Books	Script	Books	Script
356,284	150,747	92,210	22,327

7. Broken down as follows (Hebrew is not yet supported):

Arabic	50,313
Persian	5,130
Urdu	516

8. Library Automated Systems. Internet Library for Librarians, www.itcompany.com/inforetriever/sys.htm.

9. http://encompass.endinfosys.com/. Also, Endeavor Information Systems and LexisNexis announced an agreement that will make select LexisNexis databases available through ENCompass, Endeavor's digital management, organization, and linking tool. Through an XML Gateway, ENCompass will be able to search specified content from three LexisNexis research databases: LexisNexis' Academic Universe, LexisNexis Congressional Universe, and LexisNexis Statistical Universe.

10. "How to enable right-to-left language support for Word on Windows Server 2003, Windows XP, or Windows 2000." http://support.microsoft.com/default.aspx?scid=kb;[LN];Q311015. "A step-by-step article discusses the installation and configuration requirements and explains how to type in, edit, and proof (check the spelling and grammar) right-to-left languages in Microsoft Office Word 2003 or Microsoft Word 2002 on computers that are running Microsoft Windows Server 2003, Microsoft Windows XP, or Microsoft Windows 2000."

11. Website developers make sure the proper encoding is specified in the header tags. Added language tags would help the search engine and browser parse the language.

12. Some document formats allow metadata to be incorporated into documents or records, such as HTML meta-tags and Dublin Core tags, MP3 file format, Microsoft Office Properties, Adobe XAP data, and database keyword fields.

13. *Examples:* Academic Islamic Studies and Middle East, Central Asian, and Other Area Studies Sites: Academic Islamic Studies Sites and Professors of Islamic Studies, www.arches.uga.edu/~godlas/MESCenters.html. Middle East Network Information Center (MENIC) of the University of Texas at http://menic.utexas.edu.

14. The History E-Book Project, www.historyebook.org/intro.html.

15. ENALIB. http://ssgdoc.bibliothek.uni-halle.de/vlib/html/docs/StPetersburg.html.

16. ACIS for the Middle East. http://oacis.library.yale.edu/oacis/scripts/pgsearch.php?searchmod=b&ltype=en.

17. Middle East Microform Project (MEMP). www.crl.edu/areastudies/MEMP/index.htm.

18. *American Overseas Digital Library Online Union Catalog.* aodl.lib.utah.edu/ipac-cgi/ipac.

19. "The Use of the Internet for Political Action by Non-State Dissident Actors in the Middle East" by W. Sean McLaughlin, www.firstmonday.dk/issues/issue8_11/mclaughlin/.

20. http://arabworld.nitle.org/index.php.

21. *The Arab Human Development Report 2002: Creating Opportunities for Future Generations.* New York: United Nations Development Programme, Regional Bureau for Arab States, 2002.

BIBLIOGRAPHY

Aliprand, J. M. "The Unicode Standard: An Overview with Emphasis on Bidirectionality." Pp. 95–112 in *Multi-script, Multilingual and Multi-character Issues for the Online Environment,* edited by John D. Byrum Jr. and Olivia Madison. Munich: K. G. Saur, 1998.

"How to Enable Right-to-Left Language Support for Word on Windows Server 2003, Windows XP, or Windows 2000." http://support.microsoft.com/default.aspx?scid=kb;[LN];Q311015 (accessed September 14, 2004).

Madhany, H. N., *Arabicizing Windows: Enabling Windows Applications to Read and Write Arabic,* 2003. www.nclrc.org/inst-arabic3.pdf (accessed September 18, 2004).

MARBI Discussion Paper, Multilingual Authority Records in the MARC21 Authority Format. 2001-DP05. lcweb.loc.gov/marc/marbi/2001/2001-dp05.html (accessed September 18, 2004).

OCLC Arabic—Quick Reference. OCLC. September 2002. www.oclc.org/oclc/arabic/quickreference/ (accessed September 15, 2004).

RLIN Cataloguing Guide. RLG. Sept. 2002. www.rlg.org/catguide/catguide.html. (accessed September 15, 2004).

Tull, L., and D. Straley. "Unicode: Support for Multiple Languages at the Ohio State University Libraries." *Library Hi Tech* 21: no. 4 (2003): 440–50.

Unicode Character Set. www.unicode.org/ (accessed September 16, 2004).

USMARC Specifications for Record Structure, Character Sets, and Exchange Media, prepared by Network Development and MARC Standards Office, Washington, DC, Cataloging Distribution Service, Library of Congress, 1994.

Wood, Alan. *Test for Unicode Support in Web Browsers Arabic.* www.alanwood.net/unicode/arabic.html (accessed September 14, 2004).

V

BARRIERS

Chapter 15

Barriers in Introducing Information Technology in Libraries

R. N. Sharma

ABSTRACT

Libraries have been part of the world for centuries. Many changes have been introduced in libraries to meet the changing needs of users on a continuing basis. Technology was first introduced in American libraries during the second half of the twentieth century, then in the developed nations and in a few Third World countries. This chapter discusses the development of information technology (IT) with emphasis on the barriers to introducing it in Asian, African, and the Middle Eastern libraries. It will also suggest ways to reduce barriers to introducing IT.

INTRODUCTION

Libraries have played an important part in the world for centuries, storing and retrieving information for scholars and others. The method of obtaining information from libraries has changed from clay tablets and handwritten materials, to printed materials and on to microforms, CD-ROMS, and online storage, including databases on the World Wide Web. These changes were introduced in libraries to meet the changing needs of all users. During the second half of the twentieth century, technology was first introduced in the United States, followed by in other developed nations and a few Third World countries.

The late Dr. S. R. Ranganathan, an internationally known mathematician turned librarian and library educator from India, was the first person to predict in the 1950s that technology would become an important part of libraries. Since then, rapid progress has been made in the field of library and information science. The libraries have changed from manual to electronic systems in developed nations and in a few Third World countries in Asia, Africa, and the Middle East.

INFORMATION TECHNOLOGY

It was during the 1960s that more attention was paid to information science in the United States. Professors Allen Kent, James Perry, and the late Jesse Shera founded the first Center for Documentation and Communication Research at Case Western Reserve University in Cleveland, Ohio.[1] A need was felt by many researchers to have a new term to define the new developments in the United States. The technology came to be understood as *information* technology. Thus, the stage was set for a change in the field in the early 1970s. "The field was shifting from its emphasis on mainframe based, large centralized computers using batch processing to the more efficient online, interactive modes. Hardware became much smaller and less expensive, while at the same time becoming more powerful."[2]

Due to the introduction of technology, the card catalog has been replaced by the online catalog in practically all libraries in the Western world and in many countries elsewhere. The fully integrated online system has improved the operation of acquisitions, cataloging, periodicals, circulation, reference, interlibrary loans, and other departments. The CD-ROM technology and the online Internet on the World Wide Web have taken over many printed indexes and journals. The full-text journal electronic databases have brought a revolution in the library world for all researchers and scholars. The local, state, regional, and national networking in many countries has helped the users to find much-needed research material much faster. The information is now available twenty-four hours a day, 365 days a year from homes, libraries, offices, dorms, and other places for interested users. The wireless technology has further helped the users to find information quickly. The emerging digital library is making this information process much easier for all users. Introduction of e-mail and fax has also brought new life to all types of libraries. Both these new tools have become an important part of libraries. There are at present "over 600 million e-mail users sending out 30 billion e-mails a day and soon, there will be an expected 1 billion e-mail users!!! . . . By the year 2005, video e-mail . . . will replace text messages as online communications mechanism."[3] Therefore, it can be said that information technology has certainly helped to enlarge the role, capabilities, and importance of libraries in some countries in the twenty-first century.

"The library online catalog, now web-based, provides a gateway for some of the scholarly resources owned by universities and colleges. More recently it appears as a link on the library's home page."[4] It can be said that "the application of new technologies has broken down many of the barriers that existed in the past for a more extended research to global scholarly resources and services."[5] E-books have also contributed in enriching scholars. It is a growing industry. According to the American Association of Publishers, "Sales of E-books for April 2003 were up by 268.3 percent with a sales total of $900,000 . . . [but] there is near universal agreement that reading an e-book on a standard computer screen while connected to the Internet is not the reading experience users seek."[6]

In addition to the above-mentioned improvements in the libraries, due to the introduction of technology, the Library of Congress of the United States started a cooperative, digital, reference service in 2001, with the help of OCLC and sixteen other libraries. By December 2002, over three hundred libraries of various types from Asia, Canada, Europe, Latin America, and the United States became active partners in this project. This service is also known as the virtual reference with "Question Point." Though only 40 percent of its membership is from outside the United States, it is a kind of global reference network to help scholars and others in their research and information needs.[7] Looking at the rapid progress and impact of technology on libraries during the last ten years, it can be said that academic and research libraries have started giving more global access to their users through the Internet, rather than the stored material within the four walls of libraries.[8] Libraries have also started using Radio Frequency Identification (RFID) technology to protect their materials from theft and to aid in theft detection. This new system uses a fourteen-digit bar code imbedded in a chip. It replaces the magnetic bar codes, which are scanned by hand, and is scanned by machines like the ones used in large retail stores in the United States.

Libraries have certainly changed and are changing, especially in the Western world.

> Once passive storehouses, they have in some cases become active agents of social change and early adopters of new information and communication technologies. The range of materials and media they handle has diversified enormously in the last decade. Access to full-text databases, networked resources, and multimedia information systems has become the norm in a matter of years . . . due to the growth of the Internet and the World Wide Web.[9]

At least for the libraries of Western nations, "the next few years promise even greater advances—global digital libraries, intelligent interfaces, interactive books, collaboratories, intelligent agents, [a] virtual reality."[10]

Many of the advances discussed above are the result of the entrance of the "commercial sector into the library automation market."[11] At present, there are many commercial vendors in the United States and Europe who are dominating the field of library and information service. They include EBSCO, OCLC, Elsevier, H. W. Wilson, Gale, VTLS, LexisNexis, Ovid, SIRS/Mandarin, SIRSI, and Endeavor.

OCLC has 67 million records in 458 languages in its "WorldCat" with 53,000 libraries from 108 countries, but only 10,500 libraries are members of OCLC from outside the United States.[12] EBSCO, Google, Elsevier, and other databases have thousands of full-text articles in their databases for scholars and researchers. According to Bob Seal, a former president of the OCLC Members Council and University Librarian, Texas Christian University in Fort Worth, Texas, "Our users have a tremendous number of information options these days, only one of which is libraries."[13] There are many other sources where users can obtain the much-needed unlimited information twenty-four hours a day, but we must keep in mind that, in spite of the excellent progress libraries have made during the last thirty-five years due to the introduction of technology, there are still many barriers and problems that need immediate attention. Let me discuss some of the major hurdles, barriers, and problems of technology that are blocking the progress in libraries all over the world, including libraries in Asia, Africa, and the Middle Eastern countries.

BARRIERS

Funding

One of the major problems for libraries is the shortage of money for technology. The libraries were never prepared for the introduction of technology. It was dumped on libraries and librarians by smart businessmen, without any proper planning and/or training. Libraries had to divert money from other line items of the budget to introduce technology. It was very expensive, and it is still very expensive. Library budgets have been slashed even in the United States. Every library in the nation is going through a very difficult period. They cannot keep up with the rising prices of online journals and paper journals. On average, publishers and vendors have raised prices of journals between 12 percent and 15 percent every year, even if the inflation was less than 2 percent.

Budget cuts by federal and state governments for all institutions have affected library budgets also. In turn, libraries have cut their book budgets, and they have cut their journal subscriptions on paper as well as the online full-text journals and abstracts. Hardware and software are still expensive for libraries in Asian, African, and Middle Eastern countries as well as other Third World countries. They do not have enough money to buy books, have good library buildings, hire good, well-trained professional staff, or buy paper journals to support the curriculum and research needs of their students and faculty. How can they even think of buying expensive technology and train their staff in the use of technology? I have visited libraries in Asia, Africa, Latin America, and the Middle East during my library missions to various countries. In Palestine, I did not see any technology in academic libraries. In Benin, West Africa, I saw only ten computers in one university library to serve the needs of over fifteen thousand students. Those computers were for the use of graduate students only, and the limit was fifteen minutes per student. The electricity was shut off a few hours a day, during which students could not use computers. Many nations do not have enough libraries to serve the needs of their population. They do not have enough books or journals on their shelves. How can they think of technology? In a few countries elephants, donkeys, and other animals are still used to transport books to village libraries. How can they even dream of using technology in their libraries? In a few libraries of the Third World, technology has been introduced with the help of donations and grants from other countries and some help their governments, but progress is very slow and they will never be able to catch up with the libraries in the Western and developed countries.

According to Sudarsan Reghavan of the Knight Ridder newspapers, African nations owe at least $333 billion to lending institutions, with Nigeria on top of the list. She has a debt of $31.3 billion. How can these poor African nations even imagine introducing technology in their libraries? A press release by the American Library Association, dated January 12, 2005, reported that in South Asia, libraries have been hit hard by the tsunami. In Sri Lanka, 177 school libraries, fifty-three public libraries, and sixty-eight religious libraries were damaged in December 2004. The local library system was destroyed in Aceh, Indonesia. In Madras, India, the university library was damaged, and in Bangladesh, the Chittagong Public Library also suffered damage. Under these circumstances, how can libraries think of technology in libraries? They do have to rebuild their libraries with the help of other nations, take care of their people first before investing in libraries at all. In Iraq, many libraries have been destroyed by the American-led invasion of the country, and there is not enough money to rebuild them, develop collections, or introduce technology.

Illiteracy

One of the biggest barriers faced by the Asian, African, and Middle Eastern countries is the rate of illiteracy. How can they even think of introducing technology in their libraries when they have to fight poverty and illiteracy? According to the UNESCO Institute for Statistics in Bangladesh, 51.6 percent of the people between the ages of fifteen and twenty-four are still illiterate. In Benin, Africa, 46.9 percent of the people are illiterate. In South Africa 70 percent of them are illiterate, and in Iraq 61 percent of the people are illiterate. In Afghanistan, 95 percent of the population of women cannot read or write.[15] In sub-Saharan Africa, 48.4 percent are illiterate, including 83 million (58.4 percent) females. In South and West Asia, 44.7 percent of the people are illiterate, including 56.4 percent who are females—and that is 253 million females. In Arab and North Africa, 39.9 percent of the people, including 43 million women, are illiterate. In Africa, 40.2 percent, including 49.2 percent of women (113 million) are illiterate, and in Asia, 24.4 percent of people, including 32.1 percent who are females (404 million), are still illiterate. For your information, in less developed countries, 26.4 percent of the population, including 539 million women (33.9 percent) are illiterate, and 20.3 percent of the world population between the ages of fifteen and twenty-four (the ripe age for higher education), including 520 million (25.8 percent) women, are still illiterate.[16] And "women still make up two thirds of the world's adult illiterate. . . ." In Africa 49.2 percent of women are still illiterate. In South and West Asia, 56.4 percent and in Arab states and North Africa 52.2 percent of women are illiterate.[17]

Software for Computers

Another major barrier in introducing technology in libraries is that the software used for computers is not available except in English and a few other major languages, including French, German, Spanish, Chinese, Japanese, Russian, and Italian. At present, there are 186 languages spoken in 192 countries of the world. The English language leads the world, with over two billion Web pages (214,250,996), followed by Japanese (18,335,739), German (18,069,744), Chinese (12,113,803), French (9,262,663), Spanish (7,573,064), and Russian (5,900,956). Almost seven in ten (68.39 percent) of Web pages are in English, 5.85 percent in Japanese, 5.77 percent in German, 3.87 percent in Chinese, 2.96 percent in French, and 2.42 percent in Spanish.[18] It is also interesting to note that for every 1.5 English-speaking people, there is one Web page. There is only one Web page for 3.7 Swedish people, one Web page for every 43.8 Spanish-speaking people, and one Web page for 1,583.5 Arabic people.[19] There are only eighteen computers per thousand people in the Arab world, compared with a global average of seventy-eight per thousand, and only 1.6 percent of the region's population has Internet access, compared to 79 percent in the United States.[20] This important information clearly shows that the English language is dominating the Internet and the technology. There are many people and languages that are not represented on the Internet and are far behind in the progress in technology in libraries. They have very little or no information about the technology.

Many libraries in Asia, Africa, and the Middle East do not even have books; how can they introduce technology? "No more than 10,000 books were translated into Arabic over the entire past millennium, equivalent to the number translated into Spanish each year."[21] They need more books, more journals on paper rather than investing in technology, which for many countries, including Arab nations, is very expensive. They need peace in their countries so that they can send their children to schools and colleges, to learn and use libraries rather than worrying about introducing technology at this difficult time.

> Planning for technology is often complicated by the increasing speed of change such as computer technologies . . . and the introduction of new databases. . . needed by students and faculty [and other users]. . . . [F]urthermore, those new needed databases did not exist when the budget was developed and bring additional initial and recurring cost.[22]

It adds to the training costs for professional librarians and library staff regardless of which department of the library they work in—reference, circulation, interlibrary loan, library instruction, cataloging, archives, or special collections. The vendors make minor or major changes every year, making it difficult to keep up with those changes and new prices of products and services. At present, there is no dialog and/or communication between the librarians and vendors of technology and its products or input from librarians, especially in Asian, African, and Middle Eastern countries.

Technology has its limitations, too, and this in itself is another barrier in implementing IT in libraries.

> It is limited by the availability of reliable and affordable information and communication technologies. It is limited to those scholars and students who are affiliated with organizations which have the money and skills to provide access. It is limited to those who are literate, information literate and have a command of the major languages of commerce and scholarship (English in particular).[23]

There is another problem with technology that we cannot ignore.

> If I put a book [or a journal] in a room and close the door [and] open the door in 500 years, the information contained in the book will still be available. If I do that for any electronic storage device we now know, the same will not be

true, not even perhaps in ten years. The information may still be in electronic form, but we are unlikely to be able to read it with our newer technology.[24]

Online, full-text journals are attractive to many libraries, but they are very expensive, as mentioned earlier. In addition, "contractual and other bounds imposed by vendors exclude many potential users."[25] First of all, you have to buy the whole package—you have no choice in selecting journals. Many of them may not be good for the research needs of a particular library. That means you have to pay for online journals that you do not need. Secondly, you have access to those journals as long as you subscribe to the database. If you cancel your subscription to the packet, you lose access. On the other hand, if you subscribe to periodicals on paper, they become your property even when you cancel the subscription. A few vendors and publishers, including Gale and JSTOR, have included in their licenses perpetual access to subscription journals for their subscribers, but it will take a long time before it will become a reality with all publishers and vendors. This is another barrier that is keeping many libraries from subscribing to many databases through technology. Similarly, "the control obtained over scholarly intellectual property by major publishers acts as a significant barrier to widespread access. It limits the use of interlibrary loan, copying for students use and republication."[26] In addition, "Education and skills have also become more important especially computer skills and information literacy. Without those competencies, the system can only be used ineffectively and the student and researcher is unable to access the literature."[27]

I must agree with Alex Byrne that

> without the basic infrastructure of reliable electricity supplies, physical access to set up services and suitable buildings (universities, libraries, telecentres, school) in which to establish them, it is very difficult to provide or to access digital services. This is an immediate problem in [many countries of Asia, Africa, the Middle East including Afghanistan, Iraq, East Timor, and Palestine] devastated by conflict, but also in poor and remote areas of many countries.[28]

To introduce technology in libraries, all countries must have an excellent infrastructure for telecommunications, and it should be available at an affordable cost. But unfortunately, many poor countries of the world are not in a position to have international access for telecommunications because it is very expensive. In Western nations, users have unlimited access to all types of services, but in many poor countries, users still do not have access to basic HTML, or text-based services.[29] Therefore, libraries cannot introduce much-needed technology for the benefit of their users.

SUGGESTIONS

In order to have access to databases through technology, it is necessary to have up-to-date and efficient software. For many libraries,

> proprietary software may be too expensive and lack compatibility with other systems. Open source software may not be readily available. In addition, software may be unsuitable for national or local conditions because it implements a North American or European solutions and fails to accommodate local considerations.[30]

As mentioned earlier, the software may not be available in many languages except English and a few other major languages. At present, only "some library systems and other software do provide user interfaces in a variety of languages."[31] The bottom line is that technology for libraries is still very expensive, and a majority of the countries cannot afford to even think of introducing it to libraries. It is not their first or top priority. To them, fighting poverty, hunger, and literacy are the major problems. They will devote time, energy, and money to libraries and technology only after solving their major problems.

It is a well-known fact that the English language has the upper hand in the world in the use of the Internet. According to the recent statistics available, 230.6 million English-speaking people have access to the Internet, which is the gateway to scholarly research, for all users. Another 224.1 million educated people who know European languages use the Internet, compared to only 179.4 million people who know Asian languages.[32] It clearly shows that Asian nations need to train more people in information literacy and information technology through libraries. The situation in Africa and the Middle East is worse than that in Asia. Therefore, even people in those continents need to learn more about information technology and be information literate.

We are aware of the fact that "the current trend in the information era marginalizes the have-nots more and more and will [continue] to widen the gap,"[33] Therefore, it is urgent and important that the software for databases and other aspects of technology for use in libraries be developed in national languages of all countries in the world to achieve each nation's maximum benefit from the technology. "It is a task that must be seriously considered in order to safeguard against intrusive foreign cultures, particularly when they affect the young mind, and more so, national cultural identity."[34]

The literacy rate in the world has to improve before the introduction of technology can be successful. It is very important for the federal governments of all countries to invest in education and technology. It should be the top priority of all national leaders to allocate more money for education at all levels. More schools, colleges, and universities should be opened in the Asian, African, and the Middle Eastern countries, with good libraries, and the standard of education must be very high. Budgets of all institutions of higher learning must be improved. Libraries must be given adequate budgets for technology; otherwise the mission of introducing technology will not succeed. It is equally important to allocate enough money for the training in technology for librarians as well as users. Similarly, funds for continuing education for libraries and staff must be included in the library budgets on a regular basis. It is the responsibility of rich nations to help poor nations to improve their libraries and introduce technology with equality in mind. Maybe the United Nations, with the help of rich Western nations, should take the lead and help to distribute the much-needed funds for hiring staff and professionals, for their training, software, hardware, and other aspects of technology in the libraries. Grants and loans should be given to poor nations to improve their education systems with high standards. It will help them to rise to the occasion to upgrade their libraries, and join the high-speed super highway of technology. "Computer literacy and language are the most important tools needed for information literacy."[35] Therefore, it is very important that the responsibility for decision making be given to librarians rather than to business professionals, because they know what is needed to support the libraries. Business owners are more interested in huge profits, because they have become very greedy.

Individual institutions can also help libraries in poor nations to introduce technology and give training to their librarians in modern librarianship with emphasis on technology. The West Virginia State University (WVSU) Library in Institute, West Virginia, helped the National University of Benin in West Africa to improve their library and introduce technology for the benefit of students and faculty. Under my direction, the WVSU Library received a grant in the amount of $250,000 in 1999 from the United Negro College Fund and United States International Development Agency (USAID). Five librarians from the National University of Benin visited WVSU as interns, and they were prepared to meet the challenges of technology in the twenty-first century. WVSU bought thirty-two brand-new computers for them with French software, because French is the official language of Benin. A few printers, copiers, and a fax machine were also bought for them. In addition, WVSU paid for new books worth $26,000 of the University of Benin's choice to support the curriculum and the research needs of the National University of Benin, paid subscriptions for thirty journals for three years, and had two training sessions in computers to faculty and librarians for two weeks.[36] It is due to the efforts and help of WVSU that the National University of Benin Library now has an online catalog and their students have unlimited access to the Internet. It is an excellent example of helping a library. In my view, if libraries of rich nations followed the example of WVSU, we could certainly make a dent in introducing technology in all libraries in the near future. This type of cooperation is needed in Asia, Africa, and the Middle East. "Many people in these countries do not have access to even conventional telephone service."[37]

About 3.3 billion people—54.5 percent of the world's population—live in Asia. Over two-thirds of the world's poor people live in Asia, and Asia's 620 million population of illiterates is about 70 percent of the world's total population.[38] Therefore, as mentioned earlier, attention must be paid to improving the literacy rate in Asia, Africa, and the Middle East before introducing technology in their libraries. The United Nations and federal governments of all countries in the region must join hands to improve literacy rates as soon as possible. It is necessary to have a firm commitment from all parties to act on this important and urgent problem. "All action should focus on equal access to information, which is the right of citizens."[39] To bring uniformity in the world, it is very important—rather, necessary—to develop "information literacy standards . . . [because information illiteracy is] a significant obstacle throughout the world."[40] The combination of computer and information literacy in national languages and higher education with high standards for the citizens of every country will help us to achieve the goal of introducing information technologies successfully in all libraries and cross a few major barriers en route to complete success.

It must be mentioned that

> it [is] . . . unfortunate that the technological challenge for libraries arose as libraries [are] . . . experiencing the worst period of inflation and constrained resources in memory. There could hardly have been a more frustrating time in the modern history of libraries to have encountered the challenges and opportunities of creating a new library paradise.[41]

This is another reason that budget problems must be solved. But we should not go fully into technology, because "there is as yet no long term commitment for electronic format as there is for the printed materials. Therefore, we must solve the preservation and other problems before jumping into the fast moving wagon."[42] The communication charges in poor countries are still very high. More telephone lines and more reasonable rates will help libraries not only in Asia, Africa, or the Middle East but countries in the Third World to take full advantage of technology. Similarly, prices of software, and hardware should be controlled to make sure that they are affordable by all countries and their libraries. There is a need to establish Digital Legal Deposit Programs on the model of copyright offices in every country to record the new information and for preservation of all websites, online journals, and magazines and all other publications in the electronic format, including government documents. The British government and the New Zealand Parliament have already passed laws to preserve all scholarly materials in the electronic format.[43] Other countries must be guided to pass such laws for the benefit of all libraries and their users.

I have been a library administrator for over twenty-four years, and in my view, we may be in the electronic information age, but we are still far away from the reality of having access to all collections to support the research needs of all users because of the barriers discussed in this chapter. We are hurting, and it has affected the libraries all over the world more than anyone else in this modern age of library and information science.

Budget crises for libraries have reached a new height due to the introduction of technology, even in the United States. The dollar has lost up to 70 percent of its buying power during the past ten years because of the rising prices of books, journals, and electronic databases and other technological products used in libraries. "If the present trends continue, by the year 2026, the acquisitions budgets of our finest libraries will have only 20 percent buying power, but only 2 percent of the total information available in the world for our users . . . as compared to twenty years ago."[44] You can well imagine what will happen to the libraries of Asia, Africa, the Middle East, and other poor nations. Therefore, "if we don't try to create an infrastructure which is technically and electronically available to everyone we will have missed an important opportunity to change our society."[45] Finally, electricity is still not available in many countries of Asia, Africa, and the Middle East twenty-four hours a day. Many villages have no electricity at all, and in cities it is still very expensive. It creates many problems for libraries and their users. Therefore, major industrial powers of the world must help poor nations to generate more electricity, connect libraries, and make sure that the libraries and homes will never be without electricity.

CONCLUSION

If librarians have to succeed in their mission of fully introducing technology in all libraries in Asia, Africa, and the Middle East, they must make a commitment to work together in the new global environment of cooperation, development, and resource sharing, and they must invest wisely. It is important to set high standards for equal distribution of knowledge and resources, to complement the existing collections and services through information technology all over the world. In addition, it must be kept in mind that the technology is only a tool to information rather than a means to librarianship. "Collectively we can pursue collaborative approaches to purchasing, service development, standards and principles."[46] As a good team, concerned librarians can surmount all barriers, provided they pay attention to the above-mentioned suggestions to make libraries a better place for the benefit of all users in the twenty-first century. Otherwise, Asian, African, and Middle Eastern nations and their libraries will not be able to travel together on the new information super highway of the third millennium along with libraries of the developed nations.

NOTES

1. Charles H. Davis and James E. Rush, *Guide to Information Science* (Westport, CT: Greenwood Press, 1979), 3.

2. Edmond Sawer, "The Move to Information Online," *Bulletin of the American Society for Information Science* 14 (1988): 19.

3. "New Video E-mail Technology—Totally Web-Based Video E-mail," http://officemas@netzero.com (accessed October 23, 2003).

4. Barbara J. Dewey, "Considering Leadership and New Architecture for Digital Libraries," in *Leadership, Higher Education, and the Information Age,* ed. Carrie E. Regenstein and Barbara L. Dewey (New York: Neal-Schuman Publishers, 2003), 199.

5. Barbara J. Dewey, "Considering Leadership," 209.

6. Karen Coyle, "E-Books: It Is about Evolution, Not Revolution," *Net Connect*: *Supplement to Library Journal* (July 14, 2003): 8–12.

7. Brad Gauder, "Bringing Virtual Reference to Library Users in Canada," *OCLC News Letter* 261 (July 2003): 21.

8. C. C. Chen, "Global Digital Library: Technology Is Ready: How about Content?" http://web.simmonsedu/-chen/nit96/96-041-cchen (October 23, 2003).

9. "The World of Information," *Bulletin 2003–2005: School of Library and Information Science, Indiana University* (Bloomington: Indiana University, 2003), 1.

10. "The World of Information," 1.

11. Barbara B. Moran, *Academic Libraries: The Changing Knowledge Centres of Colleges and Universities* (Washington, DC: Association for Study of Higher Education, 1984), 5.

12. Telephone call, June 28, 2006, to OCLC.

13. Bob Murphy, "OCLC Members Council Focus on Innovation to Seek Solutions to Current, Future Challenges in Libraries," *OCLC Abstracts* 6, no. 45 (November 11, 2003): 1.

14. Carala Hayden, "President's Message: What Are Libraries For?" *American Libraries,* 34 (November 2003): 5.

15. "Photo Journal: Afghan Women's Voices," *BBC News World Edition,* http://news.bbc.co.uk/2/shared/spl/hi/south_asia/03/jamila/html/7.stm (accessed November 12, 2003).

16. "Illiteracy Rate and Illiterate Population, 15 Years and Older," UNESCO (UIS) Institute for Statistics (8 November 2003). http://portal.unesco.org/UIS/ev .php?URL_ID=5035&URL_DO=DO_TOPIC&URL. (accessed November 8, 2003).

17. "Illiteracy Rate and Illiterate Population."

18. "Percentages of Different Languages Used on the Web," http://cyberatlas .internet.com/big_picture/demographics/article/O,,5901_408521,00.html. (Internet.comCyberAtlas/vilaWeb) (accessed November 6, 2003).

19. "Web Pages and Languages, Ranked by the Number of Speakers per Web Page," www.uwm.edu/~iverson/htm/files/weblang.htm (sil.orgEthnologue Database) (accessed November 4, 2003).

20. Daniel Del Castillo, "Arab World Must Improve Higher Education Reform Its Backward Societies," http://chronicle.com/daily/2003/10/2003102103n.htm (accessed October 10, 2003).

21. Castillo, "Arab World Must Improve" (accessed October 10, 2003).

22. Carol Ou and Robert Dugan, "Keeping Me Awake at Night: Some Issues and Questions about Managing Technology," *Journal of Academic Libraries* 28 (November 2002): 405–10.

23. Alex Byrne, "Digital Libraries: Barriers or Gateways to Scholarly Information?" Paper presented at the International Association of Technological University Libraries Conference, Ankara, Turkey, June 2003.

24. Peter S. Graham, "Long Term Intellectual Preservation," in *Going Digital: Strategies for Access, Preservation, and Conversion of Collections to a Digital Format,* ed. Donald L. Dewitt (New York: Haworth Press, 1998), 87–88.

25. Byrne, "Digital Libraries," 3.

26. Byrne, "Digital Libraries," 6.

27. Byrne, "Digital Libraries," 7–8.

28. Byrne, "Digital Libraries," 3.

29. Byrne, "Digital Libraries," 3.

30. Byrne, "Digital Libraries," 4.

31. Byrne, "Digital Libraries," 3.

32. Byrne, "Digital Libraries," 5.

33. Anand Panyarachun, "Reaching the Information Gateways: An Unfinished Task." Paper presented at the 65th International Federation of Library Associations and Institutions Conference in Bangkok, Thailand, August 1999.

34. Panyarachun, "Reaching the Information Gateways," 10.

35. Panyarachun, "Reaching the Information Gateways," 10.

36. R. N. Sharma and Jeannie Bess, "West Virginia to West Africa and Back: An Intercontinental Collaboration," *American Libraries* 31, no. 7 (August 2000): 44–46.

37. Panyarachan, "Reaching the Information Gateways," 12.

38. Panyarachun, "Reaching the Information Gateways," 12.

39. Panyarachun, "Reaching the Information Gateways," 13.

40. Byrne, "Digital Libraries," 8.

41. Jerry D. Campbell, "Building Xanadu: Creating the New Paradise ," in *Going Digital: Strategies for Access, Preservation and Conversion of Collections to a Digital Format,* ed. Donald L. Dewitt (New York: Haworth Press, 1998), 38.

42. R. N. Sharma, "The Impact of Digital Collections on the Humanities: A Librarian's View," *Against the Grain* 15 (September 2003): 52.

43. "British Law to Preserve Electronic Publications," *LJ Academic News Wire*, 5 November 2003, 3–4.

44. Brian L. Hawkins, "Creating the Library of the Future: Incrementalism Won't Get Us There?" *Serials Librarian* 24 (1994): 18.

45. Hawkins, "Creating the Library of the Future," 27.

46. Byrne, "Digital Library," 10.

Appendix

Status of Indian Universities— Connectivity Established as of September 30, 2004

SN	University Name	Connectivity	Status
1	Alagappa Univ., Karaikudi	BB VSAT 256 Kbps	Sanctioned by UGC/PO awaited
2	Aligarh Muslim Univ., Aligarh	SCPC 512 Kbps	Commissioned on 1 Apr 2004
3	Amravati Univ., Amravati	BB VSAT 256 Kbps**	Commissioned on 14 May 2004
4	Andhra Univ., Vishakhapatnam	SCPC 512 Kbps	Commissioned on 2 Mar 2004
5	Anna Univ., Chennai	LL 2 Mbps	Commissioned on 22 May 2004
6	Annamalai Univ., Annamalainagar	SCPC 512 Kbps	Equipment shipped
7	Arunachal Univ., Itanagar	SCPC 256 Kbps**	Commissioned on 31 Oct 2003
8	Assam Univ., Silhcar	SCPC 512 Kbps	Commissioned on 27 Apr 2004
9	Avinashilingam I.H.S.& H.E.W., Coimbatore	BB VSAT 256 Kbps**	Commissioned on 29 Jan 2004
10	Awadhesh Pratap Singh Univ., Rewa	BB VSAT 256 Kbps**	Commissioned on 25 Mar 2004
11	Babasaheb Bhimrao Ambedkar Univ., Lucknow	LL 256 Kbps**	Installation in progress
12	Babasaheb Bhimrao Ambedkar Bihar University, Muzaffarpur	BB VSAT 256 Kbps**	Commissioned on 17 Sep 2004
13	Banaras Hindu University, Varanasi	LL 2 Mbps	Commissioned on 8 May 2004
14	Banasthali Vidyapith, Banasthali	SCPC 512 Kbps	Commissioned on 5 Mar 2004
15	Bangalore Univ., Bangalore	LL 512 Kbps	Commissioned on 9 Feb 2004
16	Barkatullah Univ., Bhopal	LL 512 Kbps	Installation in progress
17	Bengal Engineering College, Howrah	LL 512 Kbps	Sanctioned by UGC/PO awaited
18	Berhampur Univ., Berhampur	BB VSAT 256 Kbps**	Commissioned on 24 May 2004
19	Bharathiar Univ., Coimbatore	SCPC 512 Kbps	Equipment shipped
20	Bharathidasan Univ., Tiruchirappalli	SCPC 512 Kbps	Commissioned on 28 Apr 2004
21	Bharati Vidyapeeth, Pune	LL 512 Kbps	Commissioned on 4 Mar 2004
22	Bhavnagar Univ., Bhavnagar	BB VSAT 256 Kbps**	Commissioned on 9 Mar 2004
23	Birla Inst. of Tech. & Science, Pilani	SCPC 512 Kbps	Commissioned on 24 Mar 2004
24	Birla Inst. of Technology, Ranchi	SCPC 512 Kbps	Commissioned on 4 Mar 2004
25	Bundelkhand Univ., Jhansi	BB VSAT 256 Kbps**	Commissioned on 5 Jan 2004
26	Burdwan Univ., Burdwan	BB VSAT 256 Kbps**	Commissioned on 1 Apr 2004
27	Central Institute of English & Foreign Languages, Hydera.	LL 256 Kbps**	Commissioned on 10 Feb 2004
28	Central Inst. of Higher Tibetan Studies, Varanasi	SCPC 512 Kbps	Commissioned on 18 Sep 2004
29	Chaudhary Charan Singh Univ., Meerut	LL 256 Kbps**	Commissioned on 31 Mar 2004
30	Chhatrapati Sahuji Maharaj Univ., Kanpur	LL 256 Kbps**	Commissioned on 18 Nov 2003

31	Cochin Univ. Of Science & Tech., Cochin	LL 1 Mbps	Commissioned on 28 Jan 2004
32	Dayalbagh Edu. Inst., Agra	BB VSAT 256 Kbps	Commissioned on 3 Sep 2004
33	Deccan College Post G&R Inst., Pune	LL 512 Kbps	Commissioned on 13 May 2004
34	Deendayal Upadhyaya Uni., Gorakhpur	RL 256 Kbps**	Sanctioned by UGC/PO awaited
35	Delhi Univ., N.D.	LL 2 Mbps	Commissioned on 20 Nov 2003
36	Devi Ahilya Vishwavidyalaya, Indore	LL 512 Kbps	Commissioned on 21 Oct 2003
37	Dibrugarh Univ., Dibrugarh	SCPC 256 Kbps**	Commissioned on 13 Feb 2004
38	Dr. Babasaheb A. M. Univ., Aurangabad	SCPC 512 Kbps	Commissioned on 10 Mar 2004
39	Dr. Bhim Rao Ambedkar Univ., Agra	LL 512 Kbps	Link procurement started
40	Dr. Harisingh Gour Vishwavidhyalay, Sagar	BB VSAT 256 Kbps	PO received by ERNET
41	Dr. Ram Manohar Lohia Avadh Univ., Faizabad	BB VSAT 256 Kbps**	Commissioned on 5 Jul 2004
42	Gandhigram Rural Inst., Gandhigram	BB VSAT 256 Kbps**	Commissioned on 10 Sep 2003
43	Goa Univ.	SCPC 512 Kbps	Commissioned on 6 Feb 2004
44	Gokhale Inst. of Politics & Economics, Pune	LL 256 Kbps**	Commissioned on 27 Sep 2003
45	Gujrat Univ., Ahmedabad	SCPC 512 Kbps	Commissioned on 25 Feb 2004
46	Gujrat Vidyapith, Ahmedabad	BB VSAT 256 Kbps**	Commissioned on 16 Aug 2003
47	Gulbarga Univ.	SCPC 512 Kbps	Commissioned on 1 Jun 2004
48	Guru Ghasidas Univ., Bilaspur	BB VSAT 256 Kbps	Site survey completed
49	Guru Jambeshwar Univ., Hissar	SCPC 256 Kbps**	Commissioned on 2 May 2003
50	Guru Nanak Dev Univ., Amritsar	SCPC 512 Kbps	Commissioned on 26 Feb 2004
51	Gurukula Kangri Vishwavidyalaya, Haridwar	BB VSAT 256 Kbps**	Commissioned on 9 Sep 2003
52	Guwahati Univ., Guwahati	LL 512 Kbps	Commissioned on 27 Jul 2004
53	Hemwati Nandan Bahuguna Univ., Garhwal	SCPC 512 Kbps	Commissioned on 17 Mar 2004
54	Himachal Pradesh Univ., Shimla	SCPC 512 Kbps	Commissioned on 11 May 2004
55	Indira Kala Sangeet Vishwavidyalaya, Khairagarh	BB VSAT 256 Kbps**	Commissioned on 19 May 2004
56	Guru Gobind Singh Indraprastha University	LL 2 Mbps	Sanctioned by UGC/PO awaited
57	Jadavpur Univ., Kolkata	LL 2 Mbps	Commissioned on 25 Nov 2003
58	Jai Narain Vyas Univ., Jodhpur	SCPC 512 Kbps	Site survey completed
59	Jain Vishwa Bharati Inst., Ladnun	BB VSAT 256 Kbps**	Commissioned on 28 Jan 2004
60	Jamia Hamdard Univ., N.D.	LL 2 Mbps	Commissioned on 27 Jan 2004

(continued)

SN	University Name	Connectivity	Status
61	Jamia Milia Islamia Univ., N.D.	LL 2 Mbps	Commissioned on 1 Sep 2003
62	Jawaharlal Nehru Technological Univ., Hydera.	LL 2 Mbps	Commissioned on 1 Jun 2004
63	Jiwaji Univ., Gwalior	BB VSAT 256 Kbps**	Commissioned on 17 Sep 2003
64	Jawaharlal Nehru University N.D.	LL 2 Mbps	Commissioned on 25 Nov 2003
65	Kakatiya Univ., Warangal	BB VSAT 256 Kbps**	Commissioned on 2 Sep 2003
66	Kalyani Univ., Kalyani	LL 256 Kbps**	Commissioned on 16 Jul 2004
67	Kameshwara S. Darbhanga Sanskrit Univ. Darbhanga	BB VSAT 256 Kbps**	Commissioned on 17 Apr 2004
68	Kannada Univ., Kamalpura	BB VSAT 256 Kbps**	Commissioned on 5 Sep 2003
69	Karnatak Univ.	SCPC 512 Kbps	Commissioned on 8 Mar 2004
70	Kumaun Univ., Nainital	BB VSAT 256 Kbps**	Commissioned on 13 Feb 2004
71	Kurukshetra Univ., Kurukshetra	LL 512 Kbps	Commissioned on 15 Jul 2004
72	Kuvempu Univ., Shankaraghatta	BB VSAT 256 Kbps**	Commissioned on 6 Feb 2004
73	Lalit Narayan Mithila Univ., Darbhanga	BB VSAT 256 Kbps	Site survey completed
74	Madurai Kamraj Univ., Madurai	SCPC 512 Kbps	Commissioned on 11 Mar 2004
75	Magadh Univ., BodhGaya	BB VSAT 256 Kbps	Site survey completed
76	Maharaja Sayajirao Univ. of Baroda	SCPC 512 Kbps	Commissioned on 28 Feb 2004
77	Maharshi Dayanand Saraswati University, Ajmer	BB VSAT 256 Kbps	Sanctioned by UGC/PO awaited
78	Maharshi Dayanand University, Rohtak	LL 2 Mbps	Installation in progress
79	Mahatma Gandhi Antarrashtriya Hindi V.vidya, Wardha	BB VSAT 256 Kbps	Equipment shipped
80	Mahatma Gandhi Chitrakoot Gramodaya V.vidya, Chitrakoot	BB VSAT 256 Kbps**	Commissioned on 27 Feb 2004
81	Mahatma Gandhi Kashi Vidyapeeth, Varanasi	LL 256 Kbps	Installation in progress
82	Mahatma Gandhi Univ., Kottayam	LL 1 Mbps	Commissioned on 22 Jan 2004
83	Mangalore Univ., Mangalore	SCPC 512 Kbps	Commissioned on 1 Mar 2004
84	Manipur Univ., Canchipur, Imphal	SCPC 256 Kbps**	Commissioned on 16 Dec 2003
85	Manonmaniam Sundaranar Univ., Tirunelveli	BB VSAT 256 Kbps**	Commissioned on 1 Sep 2003
86	Maulana Azad National Urdu Univ., Hydera.	LL 256 Kbps**	Commissioned on 5 Feb 2004
87	Mizoram Univ., Aizawl	BB VSAT 256 Kbps**	Installation in progress
88	MJP Rohilkhand Univ., Bareilly	BB VSAT 256 Kbps	Site survey started
89	Mohan Lal Sukhaina Univ., Udaipur	SCPC 512 Kbps	Commissioned on 4 Mar 2004
90	Mother Teresa Women's Univ., Kodaikanal	BB VSAT 256 Kbps**	Commissioned on 16 Sep 2003

No.	University	Link	Status
91	Nagaland Univ., Kohima	BB VSAT 256 Kbps**	Commissioned on 6 Sep 2004
92	Nagarjuna Univ., Guntur	SCPC 512 Kbps	Commissioned on 1 Mar 2004
93	Nagpur Univ., Nagpur	SCPC 512 Kbps	Commissioned on 29 Mar 2004
94	National Law School of India Univ., Bangalore	LL 256 Kbps**	Commissioned on 16 Feb 2004
95	North Eastern Hill Univ.	SCPC 512 Kbps	Commissioned on 30 Sep 2003
96	North Gujrat Univ.	SCPC 512 Kbps	Commissioned on 28 Feb 2004
97	North Maharashtra Univ., Jalgaon	SCPC 512 Kbps	Commissioned on 27 Feb 2004
98	Osmania Univ., Hydera.	LL 2 Mbps	Commissioned on 10 Sep 2003
99	Patna Univ., Patna	BB VSAT 256 Kbps	Sanctioned by UGC/PO awaited
100	Pondichery Univ., Pondichery	BB VSAT 256 Kbps**	Commissioned on 1 Jun 2004
101	Potti Sreeramulu Telugu Univ., Hydera.	LL 256 Kbps**	Commissioned on 25 Nov 2003
102	Pt. Ravishankar Shukla Univ., Raipur	SCPC 512 Kbps	Commissioned on 27 Mar 2004
103	Punjab Univ., Chandigarh	LL 2 Mbps	Commissioned on 29 Dec 2003
104	Punjabi Univ., Patiala	BB VSAT 256 Kbps**	Commissioned on 2 Apr 2004
105	V B S Purvanchal University, Jaunpur	BB VSAT 256 Kbps**	Sanctioned by UGC/PO awaited
106	Rabindra Bharati Univ., Calcutta	LL 512 Kbps	Commissioned on 13 Apr 2004
107	Rajasthan Vidyapith, Udaipur	BB VSAT 256 Kbps	Sanctioned by UGC/PO awaited
108	Ranchi Univ., Ranchi	SCPC 512 Kbps	Sanctioned by UGC/PO awaited
109	Rani Durgavati Vishwavidyalaya, Jabalpur	BB VSAT 256 Kbps**	Commissioned on 25 Feb 2004
110	Rashtriya Sanskrit Vidyapeeth, Tirupati	BB VSAT 256 Kbps**	Commissioned on 18 Nov 2003
111	Sambalpur Univ., Sambalpur	BB VSAT 256 Kbps**	Commissioned on 2 Aug 2004
112	Sampurnanand Sanskrit V.vidhyalaya, Varanasi	LL 256 Kbps	Link procurement started
113	Sardar Patel Univ.	SCPC 512 Kbps	Commissioned on 25 Feb 2004
114	Saurashtra Univ., Rajkot	SCPC 512 Kbps	Commissioned on 27 Feb 2004
115	Shivaji Univ., Kolhapur	SCPC 512 Kbps	Commissioned on 3 Mar 2004
116	Shree Venkateshwara Univ., Tirupati	SCPC 512 Kbps	Commissioned on 27 Feb 2004
117	Shri Lal Bahadur Shatri Rashtriya Sanskrit Vidyapeeth,	LL 256 Kbps	Commissioned on 15 Sep 2004
118	Shri Padmavati Mahila V.vidhya, Tirupati	BB VSAT 256 Kbps**	Commissioned on 30 Aug 2003
119	SNDT Women's University, Mumbai	LL 2 Mbps	Link procurement started
120	South Gujrat Univ., Surat	SCPC 512 Kbps	Commissioned on 16 Apr 2004

(continued)

SN	University Name	Connectivity	Status
121	Sri Chandresekharendra S. Viswa Mahavidyalaya, Kancheepuram	BB VSAT 256 Kbps**	Sanctioned by UGC/PO awaited
122	Sri Jagannath Sanskrit Visvadyalaya, Puri	BB VSAT 256 Kbps	Sanctioned by UGC/PO awaited
123	Sri Krishnadevaraya Univ., Anantapur	BB VSAT 256 Kbps	PO received by ERNET
124	Sri Sathya Sai Inst. of Higher Learning , Prasanthinilayam	BB VSAT 256 Kbps**	Commissioned on 24 Mar 2004
125	Swami Raman & Teerth Marathwada Univ., Nanded	SCPC 512 Kbps	Commissioned on 18 Mar 2004
126	Tamil Univ., Thanjavur	BB VSAT 256 Kbps**	Commissioned on 3 Feb 2004
127	Tata Inst. of Social Science, Mumbai	LL 2 Mbps	Commissioned on 2 Jan 2004
128	Tezpur Univ., Tezpur	SCPC 256 Kbps**	Commissioned on 13 Jan 2004
129	Thapar Inst. of Eng. & Tech., Patiala	SCPC 512 Kbps	Commissioned on 3 Mar 2004
130	Tilak Maharashtra Vidyapeeth, Pune	LL 256 Kbps**	Commissioned on 11 Mar 2004
131	Tilka Manjhi Bhagalpur Univ., Bhagalpur	BB VSAT 256 Kbps**	Commissioned on 5 Jan 2004
132	Tripura Univ.,Agartala	SCPC 256 Kbps**	Commissioned on 12 May 2004
133	Univ. of Allahabad, Allahabad	LL 2 Mbps	Commissioned on 1 May 2004
134	Univ. of Calcutta, Calcutta	LL 2 Mbps	Commissioned on 25 Jul 2003
135	Univ. of Calicut, Kozhikode	LL 1 Mbps	Commissioned on 20 Feb 2004
136	Univ. of Hydera., Hydera.	LL 2 Mbps	Commissioned on 1 Jan 2000
137	Univ. of Jammu, Jammu	SCPC 1 Mbps	Commissioned on 15 Apr 2004
138	Univ. of Kashmir	SCPC 1 Mbps	Commissioned on 30 Aug 2003
139	Univ. of Kerela, Thiruvananthapuram	LL 1 Mbps	Commissioned on 20 Jan 2004
140	Univ. of Lucknow	LL 2 Mbps	Commissioned on 4 Jun 2004
141	Univ. of Madras, Chennai	LL 2 Mbps	Commissioned on 27 Apr 2004
142	Univ. of Mumbai, Mumbai	LL 2 Mbps	Commissioned on 27 Jan 2004
143	Univ. of Mysore, Mysore	LL 512 Kbps	Installation in progress
144	Univ. of North Bengal	SCPC 512 Kbps	Commissioned on 13 Mar 2004
145	Univ. of Pune, Pune	LL 2 Mbps	Commissioned on 12 Nov 2003
146	Univ. of Rajasthan	LL 2 Mbps	Commissioned on 17 May 2004
147	Utkal Univ., Bhubaneshwar	LL 2 Mbps	Installation in progress
148	Vidyasagar Univ., Midnapore	BB VSAT 256 Kbps**	Commissioned on 24 Dec 2003
149	Vikram Univ., Ujjain	LL 512 Kbps	Link procurement started
150	Vishva Bharati	SCPC 512 Kbps	Commissioned on 30 May 2004

Note: ** Increased to 256 Kbps on 20 Sep 04

Index

About the Contributors

C. C. Aguolu is professor of library and information science at the University of Maiduguri, Nigeria, Africa. He holds a bachelor's (honors) in classics from the University of London, England, a master's in library science from the University of Washington, United States; a master of arts in education; and a PhD in library and information science from the University of California, Berkeley. He was dean of faculty of education at the University of Maiduguri, Nigeria, from 1979 to 1982 and from 1984 to 1986. He has won many awards and honors and was elected a fellow of the International Biographical Association of England in recognition of his bibliographical work on Africa. Aguolo is the author of eight books and over sixty articles.

I. E. Aguolu is deputy university librarian and lecturer in general studies, University of Maiduguri, Nigeria. She has a bachelor's (honors) in French; a postgraduate diploma in library science; a master's in library science from the University of Ibadan; and a master of business administration from the University of Maidguri, Nigeria. She served in various positions in the library before assuming her duties as deputy university librarian in 1999. She has published over twenty articles in national and international journals and is the coauthor of *Libraries and Information Management in Nigeria* (2002).

Mohammed M. Aman is dean emeritus and professor at the University of Wisconsin–Milwaukee's School of Information Studies. He was dean of the UWM-SOIS for twenty-five years, dean of the Palmer School of Library and Information Science at Long Island University for three years, and chairman and then director of the Division of Library and Information Science at St. John's University in New York for three years. Aman began his college teaching career at Pratt Institute in New York in 1968 and continues to be active in teaching, research, consulting, and speaking. He is the author of more than fifteen books and one hundred articles, book chapters, and technical reports and has held important positions in the American Library Association, among them the chair of the International Relations Committee, International Relations Round Table. He has received the Wisconsin Librarian of the Year Award, the John Ames Humphry/OCLC/ Forest Press Award for Outstanding Contributions to International Librarianship from the American Library Association, Leadership and Appreciation Awards from the Trejo Foster Foundation, Kaula Gold Medal and Citation Award from India, awards from the Black Caucus of the American Library Association and the University of Wisconsin-Milwaukee African American Faculty and Staff Association, and the Service Award from the Association for Library and Information Science Education. Aman is the editor in chief of the *Digest of Middle East Studies* and is listed in many biographical reference books, including *Who's Who in America, Who's Who in the World, Who's Who in Library & Information Science,* and *Who's Who in American Education and Research.* He received his PhD from the University of Pittsburgh; his master's in library science from Columbia University; and his bachelor's (honors) from Cairo University, Egypt.

J. J. Britz has been the dean of the School of Information Studies at University of Wisconsin, Milwakee since 2004. He was a professor in the Department of Information Science at the University of Pretoria, South Africa and also a visiting professor in the Department of Information Science at the University of Amsterdam in the Netherlands. He holds a doctoral degree in information science and Christian ethics from the University of Pretoria, South Africa, and specializes in the fields of information policy, information ethics, and economics of information. He is the author of sixty professional and research papers. Britz is involved in a project on the measuring of information poverty, and he serves on a number of international bodies in the field of information ethics.

Ching-chih Chen is a professor at the Graduate School of Library and Information Science at Simmons College, Boston. She has been a speaker in over forty countries and author and editor of over thirty-five books and more than 180 scholarly journal articles. She produced the award-winning interactive videodisc and multimedia CD entitled *The First Emperor of*

China and was the chief conference organizer of a series of twelve international conferences on New Information Technology from 1986 to 2001.

Since 1993, she has been advocating the global digital library concept by linking libraries, museums, and archives. Since 2000, Chen has led a NSF/International Digital Library Project, Chinese Memory Net (CMNet). She is also co-PI with Professor Raj Reddy of the China-US Million Book Digital Library Project, and is a member of the Advisory Committee of DE-LOS (European Digital Library Network) and co-chaired the DELOS/NSF Working Group on Digital Imagery for Significant Historical, Cultural, and Heritage Materials. She has been addressing the need for an international consortium in making cultural and heritage digital contents accessible to users.

A fellow of the American Association for Advancement of science, Chen has received many awards and honors, including the Best Information Science Teacher Award of the American Society for Information Science, the Library and Information Technology Association's (LITA) Library HiTech Award, the LITA/Gaylord Award for the Advancement in Library and Information Technology, and the LITA/Frederick G. Kilgour/Online Computer Library Center/American Library Association Award for Research in Library and Information Technology for 2006. During 1997–2002, she served as a member of the U.S. President's Information Technology Advisory Committee.

In 2004, she was a keynote speaker at the International Conference on Digital Libraries in Delhi, India; the Libraries in the Digital Age (LIDA 2004) International Conference, Dubrovnik and Mljet, Croatia; the International Conference on Digital Libraries, Beijing, China; the International Asian Digital Library Conference, Shanghai, China; and the Invited Annual Lecturer of the Annual Lecture in Informatics in Bangalore, India. She received her bachelor's degree from the National Taiwan University, a master's in library science from the University of Michigan, and a PhD from Case Western Reserve University.

Rajwant S. Chilana has master's degrees in science (zoology) and library and information science from the University of Rajasthan, Jaipur, India. In 1983, he was awarded a doctorate on the topic "University Library Buildings in India: A Critical Appraisal." Chilana is an associate professor of library administration and South Asian studies librarian at the University of Illinois, Urbana-Champaign, Illinois. He worked as librarian at the University of British Columbia, Vancouver, Canada, and as a library manager of the Fraser Valley Regional Library, British Columbia, Canada. In India, he served as a senior librarian and teaching faculty at the University of Delhi and also worked as chief librarian at the Guru Nanak Institute for Comparative Study of Religions, New Delhi. He has published five books and several research papers on various aspects of South Asian studies. His latest book is *International Bibliography on Sikh Studies*. Chilana is involved in professional associations and was chair of publication committee of the AAMES/ACRL and chief editor of the AAMES *Newsletter*. He is an active member of the American Library Association, the Association for Asian Studies, the Association of College and Research Libraries, the Indian Library Association, the Society for Information Science, and the Indian Association for Special Libraries and Information Centers. He was honored with the "Punjab (India) National Librarian" Award for 2005 for his outstanding contributions in promoting librarianship in India, Canada, and the United States.

V. S. Cholin worked with the INFLIBNET Centre, India, from 1993 until his death in 2005. He received his master's degree and PhD in library and information science from Karnatak University, Dharwad, India. He was awarded the Fulbright Professional Fellowship 2004–2005, the SIS-Professional Young Scientist Award in 2004, and the Best Paper Award from the Raja Rammohun Roy Library Foundation (RRLF) for writing the best professional article in 2002. Cholin published over twenty-five papers and was head of the Informatics Division of the INFLIBNET centre and looking after the prestigious UGC-Infonet E-Journals Consortium.

Gregory A. Finnegan is an associate librarian for public services and head of reference in the Tozzer Library of Harvard University Library, as well as managing editor of *Anthropological Literature*, the bibliography of over 500,000 citations to journal articles and articles in edited volumes in anthropology. Finnegan has served as a general reference and instructional librarian and as a subject bibliographer. He has a bachelor's degree from Raymond College of the University of the Pacific, a PhD in anthropology from Brandeis University, and an AM from the University of Chicago's Graduate Library School. He has chaired the Anthropology and Sociology Section of the Association of College and Research Libraries and has been chair three times of the Africana Librarians Council of the African Studies Association. Finnegan has taught anthropology at Lake Forest College and Dartmouth College and served as film review editor of the *American Anthropologist*. He has done anthropological field research in Burkina Faso, Africa, and in Antigua, West Indies, along with archaeological fieldwork in California.

I. Haruna is a senior lecturer in library science at the University of Maiduguri, Nigeria, Africa. He has a bachelor of library science from Ahmadu Bello University, Zaria, and a master of library science and PhD in librarianship from the University of Ibadan, Nigeria. Before joining the University of Maiduguri, he was law librarian at the High Court of Justice in Makurdi and Lokoja, Nigeria. Haruna teaches information science, cataloging and classification, special Librarianship, and library administration. He has published over twenty articles in national and international referred journals.

Ali Houissa has been the Middle East and Islamic studies bibliographer at Cornell University Library, Ithaca, New York, since 1988. He is a graduate of the School of Library and Information Science, Indiana University, Bloomington; Fachhochschule für Bibliotheks und Dokumentationswesen, Cologne, West Germany; and the University of Tunis, Tunisia. Houissa serves on the executive board of the Middle East Librarians' Association of North America and as councilor-at-large of the American Library Association, 2003–2006.

Binh P. Le holds a MLS from Indiana University, Bloomington, and is a PhD candidate at Temple University, Philadelphia. He is an associate librarian at Pennsylvania State University, Abington. He was an assistant reference and instruction librarian at Purdue University. His publications include *A New Way to Serve the Library User: A Global Perspective* (editor) and articles appearing in *Library Administration and Management* (forthcoming), *International Third World Studies: Journal and Review*, *Journal of Asia-Pacific Affairs*, *Bulletin of Bibliography*, and *Encyclopedia of Library History*. He is also a contributor to *American Reference Books Annual* and is listed in *Who Is Who among Asian Americans*. He is the chair of the Asian, African, and Middle East Section of the Association of College and Research Libraries/American Library Association for 2006–2007.

Jing Liu studied library science at Wuhan University in China and received her MLS from the University of Washington. She worked as an assistant librarian for three years for the Chinese Academy of Social Sciences in Beijing. After years of experience with public and corporate libraries in the United States and Canada, such as New York Public Library, Jing became the Chinese studies librarian and technical services supervisor at Asian Library of the University of British Columbia (UBC), Canada.

Jing's research interests are in the digital library development in China and international collaborative virtual reference services. She has published books and articles in both English and Chinese and has presented papers at international conferences to promote international collaboration in reference and document delivery, focusing on Chinese studies. She provided her practical library experience to the Chinese Librarian Training program of the Christian Foundation for Higher Education and has received many awards from UBC and libraries in China.

P. J. Lor is secretary general for the International Federation of Library Associations and Institutions, Netherlands. He was Professor Extraordinary in the Department of Information Science at the University of Pretoria, South Africa. Lor also served as the national librarian and chief executive officer of the National Library of South Africa, university librarian at the University of Bophuthatswana, professor at the University of South Africa's Department of Library and Information Science, director of the State Library of Pretoria, and executive director of the Foundation for Library and Information Services Development. He was involved in developing policy for library and information services for postapartheid South Africa and played a leading role in the formation of the Library and Information Association of South Africa. Lor received his bachelor's and master's degrees in library science from the Universities of Stellenbosch and Pretoria, South Africa, respectively.

T. A. V. Murthy is the director of INFLIBNET and president of SIS, India. He holds BSc., MLI Sc., MSLS., and a PhD. He worked at the managerial level in a number of libraries in many prestigious institutions in India, including the National Library, University of Hyderabad, Catholic University, and Case Western Reserve University in the United States. His contributions include: KALANIDHI at IGNCA, Digital Laboratory at CIEFL. Associated with a number of universities in India, Murthy has guided a number of PhDs. He has been associated with the national and international professional associations and expert committees and has published many research papers. He has visited many countries and organized several national and international conferences and programs.

James J. Natsis is the coordinator of international affairs at West Virginia State University, Institute, West Virginia. He is a former Peace Corps volunteer who served for two years in Chad, served two years in sub-Saharan Africa, and conducted his doctoral research in North Africa. He holds a bachelor's degree in French from the University of Missouri at St. Louis, an MA in international affairs–African studies, and the PhD in international education from Ohio University. He has published a monograph and several journal articles, directed several grants, and presented papers at many regional and national conferences.

H. Kay Raseroka is director of Library Services at the University of Botswana. She held the position of president of the International Federation of Library Associations and Institutions for the period 2003–2005. She has been university librarian for over twenty years. During this period, she has been involved in facilitating organizational change from a focus on collection organization to subject librarianship that considers customer service a priority, ranging from curriculum support services to teaching information literacy skills. In preparation for the trend of rapid growth in numbers of students and faculty and the variety of academic programs offered by the university, she spearheaded advocacy for construction of a modern library with modern ICT infrastructure for enhancing learning, teaching, and research, while sustaining appropriate levels of collection development.

Raseroka has been involved in regional academic library research activities through the Standing Conference of National and University Libraries, Eastern Central and Southern Africa. She contributed to "University Libraries in Africa: A Review of Their Current State and Future Potential," published in the *Sub-Saharan Review of University Libraries* (1997). She has been deeply engaged in advocacy of libraries as facilitators of the emerging information and knowledge society, through the World Summit on the Information Society, 2003–2005.

Sherif Kamel Shaheen is head of the National Library of Egypt and a professor of library and information science at Cairo University, Egypt. He has worked in England, Germany, Saudi Arabia, and Egypt. Shaheen is the recipient of many awards and honors and has published many articles in various library journals. He received his bachelor's and PhD degrees in library and information science from Cairo University, and a master's degree in library and information studies from Leeds University, England.

Ian Yiliang Song is the digital initiatives coordinator at Simon Fraser University Library, Burnaby, British Columbia, Canada. He worked at the Vancouver Public Library in British Columbia, Canada; the University of Loughborough, United Kingdom; and Suzhou Medical College Library, Jiangsu Province, China. Song received his bachelor of medicine from Normal Bethune University of Medical Sciences, Chang-Chun, China, and master of library and information studies from the School of Library, Archives, and Information Studies at the University of British Columbia.

Andrew H. Wang is the executive director of OCLC Asia Pacific, which he founded in 1986. He also founded OCLC Conversion Services and the OCLC CJK Program, an automation system of library information in Chinese, Japanese, and Korean characters.

Before joining OCLC in 1976, he was an assistant university librarian for technical services at Denison University, Granville, Ohio, and catalog librarian at St. Mary's College of Maryland.

He received a bachelor's degree in journalism from National Cheng-Chi University, Taipei, Taiwan; a master's degree in library service from Atlanta University, Atlanta; and a master's degree in business administration from Ohio State University. He also studied in the computer science PhD program at Ohio State University.

Wang holds a lifetime membership in Beta Phi Mu, the International Honor Society of Library Science, and in 1995, he was appointed an advisor to the National Central Library, Taipei, Taiwan. He has also served on many other professional advisory boards and committees and is listed in the *Who's Who in the World* and *Who's Who in America*. Since 1986, Wang has lectured extensively throughout Asia and the Pacific Region on library automation, the electronic library, and online access to information.

About the Editor

R. N. Sharma is dean of the library at Monmouth University, West Long Branch, New Jersey. He is an active professional and has won many awards, grants, and honors, including the Academic Research Librarian of the Year from the Association of College and Research Libraries/ALA for 2005, and the Humphry/OCLC/Forest Press Award in 1997 from the American Library Association, for his significant achievements and contributions to international librarianship. He is the author/editor of ten books and has published over 250 articles, editorials, interviews, conference reports, and book reviews in various journals of Asia, Europe, and the United States. He has been the editor of *Library Times International* since 1984 and has served on many editorials boards of books and journals. He has served on over thirty committees of the American Library Association and the Association of College and Research Libraries, and he has also chaired many committees, including the Asian, African, and Middle Eastern Section of the Association of College and Research Libraries and Near East and South Asia Subcommittee of the American Library Association. Sharma was president of the Asian/Pacific American Librarians Association, has organized many conferences, and has presented twenty-seven papers at national and international conferences. He is listed in over twenty bibliographical dictionaries, including *Who's Who in the World, Contemporary Authors,* and *Who's Who in Education.*

Sharma was director of the library at the West Virginia State University Library, director of the library at the University of Evansville, head librarian at Penn State University–Beaver Campus, assistant director of

Public Services and Collection Development at the University of Wisconsin–Oshkosh, and reference librarian at Colgate University.

Sharma has a bachelor's degree (honors) and MA in history from the University of Delhi, India; a master's degree in library science from the University of North Texas; and a PhD from the State University of New York, Buffalo. He has visited Africa, Asia, Europe, Latin America, and the Middle East on many library assignments.